CW01249674

New Directions in the Philosophy of Science

Series Editor: **Steven French**, Philosophy, University of Leeds, UK

The philosophy of science is going through exciting times. New and productive relationships are being sought with the history of science. Illuminating and innovative comparisons are being developed between the philosophy of science and the philosophy of art. The role of mathematics in science is being opened up to renewed scrutiny in the light of original case studies. The philosophies of particular sciences are both drawing on and feeding into new work in metaphysics and the relationships between science, metaphysics and the philosophy of science in general are being re-examined and reconfigured.

The intention behind this new series from Palgrave Macmillan is to offer a new, dedicated, publishing forum for the kind of exciting new work in the philosophy of science that embraces novel directions and fresh perspectives.

To this end, our aim is to publish books that address issues in the philosophy of science in the light of these new developments, including those that attempt to initiate a dialogue between various perspectives, offer constructive and insightful critiques, or bring new areas of science under philosophical scrutiny.

Titles include

THE APPLICABILITY OF MATHEMATICS IN SCIENCE
Indispensability and Ontology
Sorin Bangu

PHILOSOPHY OF STEM CELL BIOLOGY
Knowledge in Flesh and Blood
Melinda Fagan

SCIENTIFIC ENQUIRY AND NATURAL KINDS
From Planets to Mallards
P.D. Magnus

COUNTERFACTUALS AND SCIENTIFIC REALISM
Michael J. Shaffer

MODELS AS MAKE-BELIEVE
Imagination, Fiction and Scientific Representation
Adam Toon

Forthcoming titles include

THE PHILOSOPHY OF EPIDEMIOLOGY
Alex Broadbent

SCIENTIFIC MODELS AND REPRESENTATION
Gabriele Contessa

CAUSATION AND ITS BASIS IN FUNDAMENTAL PHYSICS
Douglas Kutach

BETWEEN SCIENCE, METAPHYSICS AND COMMON SENSE
Matteo Morganti

ARE SPECIES REAL?
Matthew Slater

THE NATURE OF CLASSIFICATION
John S. Wilkins and Malte C. Ebach

New Directions of the Philosophy of Science
Series Standing Order ISBN 978–0-230–20210–8 (hardcover)
(*outside North America only*)

You can receive future titles in this series as they are published by placing a standing order. Please contact your bookseller or, in case of difficulty, write to us at the address below with your name and address, the title of the series and the ISBN quoted above.

Customer Services Department, Macmillan Distribution Ltd, Houndmills, Basingstoke, Hampshire RG21 6XS, England

Models as Make-Believe

Imagination, Fiction and Scientific Representation

Adam Toon
Department of Philosophy, University of Bielefeld, Germany

palgrave
macmillan

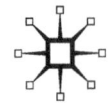

© Adam Toon 2012

All rights reserved. No reproduction, copy or transmission of this publication may be made without written permission.

No portion of this publication may be reproduced, copied or transmitted save with written permission or in accordance with the provisions of the Copyright, Designs and Patents Act 1988, or under the terms of any licence permitting limited copying issued by the Copyright Licensing Agency, Saffron House, 6–10 Kirby Street, London EC1N 8TS.

Any person who does any unauthorized act in relation to this publication may be liable to criminal prosecution and civil claims for damages.

The author has asserted his right to be identified as the author of this work in accordance with the Copyright, Designs and Patents Act 1988.

First published 2012 by
PALGRAVE MACMILLAN

Palgrave Macmillan in the UK is an imprint of Macmillan Publishers Limited, registered in England, company number 785998, of Houndmills, Basingstoke, Hampshire RG21 6XS.

Palgrave Macmillan in the US is a division of St Martin's Press LLC, 175 Fifth Avenue, New York, NY 10010.

Palgrave Macmillan is the global academic imprint of the above companies and has companies and representatives throughout the world.

Palgrave® and Macmillan® are registered trademarks in the United States, the United Kingdom, Europe and other countries.

ISBN: 978–0–230–30121–4

This book is printed on paper suitable for recycling and made from fully managed and sustained forest sources. Logging, pulping and manufacturing processes are expected to conform to the environmental regulations of the country of origin.

A catalogue record for this book is available from the British Library.

A catalog record for this book is available from the Library of Congress.

10 9 8 7 6 5 4 3 2 1
21 20 19 18 17 16 15 14 13 12

Printed and bound in Great Britain by
CPI Antony Rowe, Chippenham and Eastbourne

For Angela

Contents

List of Figures x
Series Editor's Foreword xi
Preface xiii
Acknowledgements xvi

Introduction 1

1 Models and Representation 6
   1.1 Models 6
      1.1.1 Physical models 6
      1.1.2 Theoretical modelling 7
   1.2 The ontology of theoretical modelling 10
      1.2.1 The problem: missing systems 10
      1.2.2 Description-fitting objects: abstract objects 13
      1.2.3 Description-fitting objects: fictional characters 15
   1.3 The problem of scientific representation 18
      1.3.1 The problem 18
      1.3.2 Misrepresentation 23
      1.3.3 Two kinds of solution 25
      1.3.4 Does the problem exist? 26
      1.3.5 Stipulation and salt shakers 31
   1.4 Conclusion 33

2 What Models Are 34
   2.1 Walton's theory: representation and make-believe 34
      2.1.1 Props and games 34
      2.1.2 Representations 35
   2.2 Models as make-believe 37
      2.2.1 Physical models 37
      2.2.2 Theoretical modelling 38
   2.3 Make-believe and the ontology of modelling 41
      2.3.1 Doing without description-fitting objects 41
      2.3.2 Learning about theoretical models 45
      2.3.3 Talking about theoretical models 48

	2.4	The indirect fictions view	53
		2.4.1 Direct and indirect views of theoretical modelling	54
		2.4.2 Make-believe and the indirect fictions view?	57
		2.4.3 Deferring the problem?	59
	2.5	Conclusion	60
3	How Models Represent		61
	3.1	Make-believe and model-representation	61
		3.1.1 The account	61
		3.1.2 Make-believe and salt shakers	63
		3.1.3 Make-believe, misrepresentation and realism	65
		3.1.4 Make-believe and similarity	67
	3.2	Models and works of fiction	69
	3.3	Models without objects	75
		3.3.1 Problem cases	75
		3.3.2 What is the problem?	77
		3.3.3 Existing accounts and models without objects	79
		3.3.4 Solving the problem	81
	3.4	Conclusion	82
4	Carbon in Cardboard		84
	4.1	Chemistry in space	85
		4.1.1 Isomerism and 'chemical structure'	85
		4.1.2 Optical activity and 'absolute isomers'	88
		4.1.3 Van't Hoff and the tetrahedral carbon atom	89
		4.1.4 The reception of van't Hoff's work	93
	4.2	Building molecules	95
		4.2.1 Models before van't Hoff	95
		4.2.2 Carbon in cardboard	97
		4.2.3 The role of models	100
	4.3	How chemical models represent	103
	4.4	Conclusion	107
5	Playing with Molecules		108
	5.1	Molecular models in use	109
		5.1.1 The study	109
		5.1.2 Talking about models	111

	5.1.3 Looking at models	113
	5.1.4 Manipulating models	114
	5.1.5 Play, pretence and realism	116
5.2	Molecular models as props	117
5.3	Playing with molecules	119
	5.3.1 Participation and depiction	120
	5.3.2 Manipulating molecules	121
	5.3.3 Seeing molecules	123
	5.3.4 Touching molecules	125
5.4	Imagined experiments	126
5.5	Conclusion	129
Conclusion		131
Notes		134
References		138
Index		145

Figures

1.1	Indirect views of theoretical modelling	19
2.1	Indirect views of theoretical modelling	43
2.2	Models as make-believe	44
4.1	Crum Brown's graphical formulas	87
4.2	Diagrams from van't Hoff's original pamphlet	91
4.3	Hofmann's 'glyptic formulas'	96
4.4	Kekulé's tetrahedral models	96
4.5	Van't Hoff's early models of the tetrahedral carbon atom	98
4.6	Van't Hoff's templates (face bonding models)	99
4.7	Van't Hoff's templates (vertex bonding models)	99

Series Editor's Foreword

The intention behind this series is to offer a new, dedicated publishing forum for the kind of exciting new work in the philosophy of science that embraces novel directions and fresh perspectives. To this end, our aim is to publish books that address issues in the philosophy of science in the light of these new developments, including those that attempt to initiate a dialogue between various perspectives, offer constructive and insightful critiques, or bring new areas of science under philosophical scrutiny.

Adam Toon's provocative and exciting new book fits the series remit perfectly. The nature of scientific representation has become a major topic within the philosophy of science that has attracted considerable attention, not least in part because of the possibility for fruitful cross-fertilization with the philosophy of art, where, of course, this issue has been discussed in detail. However, much of the focus so far has concerned issues and concerns associated with representation in painting, for example, and Toon takes the discussion in an entirely new direction by drawing upon the philosophy of literature and the way in which fiction is treated philosophically. He is not alone, of course, in seeing this fictionalist aspect as constituting a central feature of scientific models which, as he says, are typically taken to be true of no real object. However, rather than taking the models themselves to be abstract or fictional, Toon imports Walton's 'pretence' analysis of fictions from the philosophy of literature and develops it in this context. On this view models are regarded as 'props' in something akin to games of make believe, which represent the world by prescribing 'imaginings' about it. This offers an entirely new and thought provoking perspective on the issue of the nature and role of representation in science.

Thus, this book represents a novel approach that not only further extends and develops the interconnections between the philosophy of science and the philosophy of art, but also tackles a number of major concerns in the philosophy of science to do with the nature and role of models and modelling. Furthermore, it does so in the context of detailed case studies from chemistry, illustrating how visual and tactile forms of participation may be involved in modelling and how this supports forms of 'imagined experiments'. In this manner Toon takes the debate forward and away from the rather simplistic examples that have been employed in the literature so far, towards those features of the scientific enterprise that are more complex, more interesting and closer to actual practice.

In all these ways then – in drawing on the philosophy of literature, in offering a radically new account of representation in science and in deploying case studies that are closer to actual scientific practice than the examples usually presented – this book accomplishes precisely what we set out to do with this series.

Steven French
Professor of Philosophy of Science
University of Leeds

Preface

This book began life as a doctoral thesis in the Department of History and Philosophy of Science at the University of Cambridge. My first experience of the HPS Department was as an undergraduate when, after two years of a degree in maths and physics, I decided on a change of direction, and spent my final year studying history and philosophy of science. It proved to be an enormously exciting, rewarding and challenging year, and I feel extremely privileged to have been welcomed into such an open, friendly and stimulating intellectual environment. I am particularly grateful to my supervisor, Martin Kusch, for all of his hard work, support and encouragement throughout my PhD. I feel very lucky to have had such an extraordinarily able and conscientious supervisor for my doctoral work. Some of the most inspiring and memorable experiences of my time in the HPS Department came in Peter Lipton's wonderful lectures, and I feel extremely fortunate that I was able to receive Peter's comments on early drafts of my thesis.

Since leaving Cambridge I have been lucky enough to find a home in another exceptionally friendly, supportive and stimulating environment for philosophical study, in the Department of Philosophy at the University of Bielefeld. I would like to thank all of my colleagues in Bielefeld, especially Martin Carrier, Ulrich Krohs, Maria Kronfeldner, Johannes Lenhard, Cornelis Menke, Alex Wiebke and Torsten Wilholt, for all of their support and for making me feel so welcome. Thanks also to Torsten for very helpful comments on the first few chapters of the book, and to the students in my course on 'Representation in Science and Art', for their challenging and thought-provoking criticisms of work in the field (including my own). And thank you especially to Martin Carrier, without whose support this book would not have been possible.

Many other people have given extremely helpful suggestions, comments, criticisms and advice during my work on the book. I am particularly grateful to Martin Thomson-Jones for his generous support and encouragement for the project, and for extremely helpful and constructive criticisms of a draft of the book. His own work has done a great deal to help me understand the problems that an account of scientific modelling must solve. A special thank you to Stacie Friend, too, both for her friendship and encouragement during my PhD, and for helpful and enjoyable discussions on fiction, imagination and much else besides. Thanks also to Roman Frigg for his generous support, for lots of stimulating discussions on where we agree and disagree, and for inviting me to so many excellent events on models and fiction during my PhD, including the 'Fictionalism' reading group at LSE during 2007–8. Thanks to participants in this group for extremely interesting and helpful discussion.

My debt to Kendall Walton is considerable, and engaging with his work has taught me a huge amount, as well as being a lot of fun. I am also grateful for his kind and constructive comments on the project, although of course the responsibility for my use (or misuse) of his ideas is entirely my own. In addition, I would like to thank Nancy Cartwright, Soraya de Chadarevian, Anjan Chakravartty, Hasok Chang, Gabriele Contessa, Tim Crane, John Forrester, Ronald Giere, Peter Godfrey-Smith, Nick Hopwood, Nick Jardine, Mary Leng, Aaron Meskin, Chris Pincock, Eleanor Robson, Simon Schaffer, Jim Secord, Mark Sprevak, Mauricio Suárez, Liba Taub, Paul Teller and Michael Weisberg for very helpful discussion and correspondence. Thanks to Tim Lewens for his help and support, as well as very helpful comments and criticisms on my doctoral thesis, and to Marina Frasca-Spada, for her kind support, and for supervising an undergraduate thesis on Hume that helped to get me hooked on philosophy. And thanks also to Stephen Boulter, Mark Cain, Dan O'Brien, Constantine Sandis and my students at Oxford Brookes, for a hugely enjoyable term spent there.

I have presented some of the ideas from the book at many conferences and workshops over the past few years, including the 'Beyond Mimesis and Nominalism' conference held at LSE in June 2006, the Philosophy Workshop in Cambridge in June 2006, the CMM Graduate Conference in Leeds in June 2007, the 'Scientific Models: Semantics and Ontology' workshop in Barcelona in July 2007, the Inaugural Conference of the Centre for Literature and Philosophy at the University of Sussex in June 2008, a Departmental Seminar in the HPS Department at Cambridge in November 2008, the 'Models and Fiction' workshop at LSE in March 2009, a Departmental

Colloquium at the Leibniz Universität in Hannover in May 2010, a Departmental Seminar at Oxford Brookes University in December 2010, and the Biennial Conference for Philosophy of Science in Practice in Exeter in June 2011. I would like to thank participants at all of these events for helpful and thought-provoking questions, comments and criticism, particularly my commentator in Barcelona, Manuel García-Carpintero.

Research for this book was supported by The Arts and Humanties Research Council, the Darwin Trust of Edinburgh, the Williamson and Rausing Funds of the Department of History and Philosophy of Science, Cambridge and King's College, Cambridge. I am extremely grateful to all of these institutions for their support. Thanks also to those who took part in the empirical study in Chapter 5. And thank you to Steven French, for including the book in this series, and to Pri Gibbons at Palgrave Macmillan for her help during the writing process.

All in all, this book has been a long time in the making, and I would like to thank my friends and family, especially my mum, dad and sister, for all of their love and support over the years, as well as the wonderful teachers at Highfields School in Matlock, for making me love learning in the first place. Finally, I would like to thank my wife, Angela, for just about everything. This book is dedicated to her.

Acknowledgements

Parts of this book incorporate revised and reworked versions of material from the following articles. I am grateful to the editors and publishers involved for permission to use that material here.

The ontology of theoretical modelling: models as make-believe. *Synthese, 172*(2), 301–15.

Models as make-believe. In R. Frigg and M. Hunter (eds), *Beyond Mimesis and Convention: Representation in Art and Science* (pp. 71–96). Dordrecht: Springer.

Novel approaches to models. *Metascience, 19*(2), 285–8.

Playing with molecules. *Studies in History and Philosophy of Science, 42*, 580–9.

I would also like to thank the Museum Boerhaave, Leiden and the Museum for the History of Sciences, Ghent University for permission to use the images of van't Hoff and Kekulé's models in Chapter 4.

Introduction

One of the main things that scientists do is to represent the world. They do so in many different ways, through theories, equations, formulas, diagrams, graphs, photographs, videos, traces, sketches, watercolours, maps, X-rays and more else besides. And they offer us representations of a vast array of different things: subatomic particles, atoms, molecules, electromagnetic fields, microbes, beetles, the migration patterns of birds, dinosaurs, weather systems, world economic markets, the rise in global temperatures, tectonic plates, the movement of planets in the solar system, distant galaxies, black holes and the big bang.

This book is about how scientific representation works. Although scientists use many different representational devices, this book focuses in particular on scientific *models*. Scientists often try to understand a complex, real world phenomenon by first constructing a simplified or idealised model of it. Sometimes they might construct a *physical model*, such as the scale models that engineers build to test new structures. Often, however, scientists simply write down a set of assumptions or equations, and so come up with a *theoretical model*. A scientist might try to understand the solar system by assuming that the planets are perfect spheres subject only to the gravitational field of the sun, for example, or treat the molecules of a gas as if they were a collection of tiny billiard balls. Modelling is extremely important in the sciences. Indeed, some even argue that models are involved whenever complex, mathematical theories are applied to the world (e.g., Cartwright, 1983).

Models raise a number of important questions. Some of these questions concern scientific realism. Realists claim that successful scientific theories are true, or at least approximately true. And yet models typically make many assumptions that are false of the systems they model. The planets are not perfect spheres and molecules are not billiard balls. Even worse, scientists sometimes offer a number of different, incompatible models for the same phenomena, each of which is successful for different purposes. Models also raise questions for our understanding of laws of nature. We normally take these laws to be true in all places and at all times. And yet many of the fundamental laws we find in science are true only in the limited, highly simplified domains described by scientists' models. This also presents problems for our accounts of scientific explanation. According to one common view, we explain a phenomenon by showing how it may be deduced from the laws of nature. But if laws are true only of models, how do we explain the behaviour of the complex, messy systems we find in the real world?

This book will not attempt to address all of the problems posed by scientific models. Instead, it will focus on two key questions: what are models, and how do they represent the world?

The first question asks what models are. It is not difficult to say what physical models are: they are bits of wood or metal or plastic. Theoretical modelling is more problematic, however. In theoretical modelling, scientists often make assumptions that are true of no actual, physical object: there are no perfect spheres, and even real billiard balls do not satisfy the assumptions made in the billiard ball model of gases. And yet scientists commonly talk as if there were such objects and as if they can find out about their properties. They talk as if they were investigating a 'model-system' which satisfies the assumptions they make. Theoretical modelling therefore gives rise to a number of ontological worries: how are we to make sense of the fact that a large part of scientific practice seems to involve talking and learning about things that do not exist?

The second main question addressed in this book asks how models represent the world. Many scientific models represent objects or events in the world: the Bohr model represents the atom, Crick and Watson's famous double-helical model represents DNA, and the billiard ball model of gases could be used to represent helium, or hydrogen or oxygen. The problem of scientific representation asks how these models do this. One way to understand this problem is to compare models to pictures. In itself, it seems, a painting like

Constable's *Salisbury Cathedral from the Meadows* is merely a set of brushstrokes on a piece of canvas. And yet it represents horses pulling a cart through a stream, and Salisbury Cathedral itself beneath a rainbow. The problem of depiction asks how paintings can do this. Similarly, Crick and Watson's original DNA model was simply a collection of metal rods and plates held in place by clamps. And yet it represented the complex helical structure of the DNA molecule. How did the model do this?

This book will offer answers to these questions about scientific modelling by looking to what, at first sight, might seem an unlikely source of inspiration: children's games of make-believe. I shall argue that scientific models function like the dolls and toy trucks of children's imaginative games. In order to develop this idea, I will draw extensively on the work of the philosopher of art Kendall Walton (1990). Walton offers a sophisticated framework for understanding games of make-believe, and uses this framework to provide a general theory of art and fiction. I will draw on Walton's framework throughout this book, although my treatment of models will not rely upon some of the more controversial aspects of Walton's theory of art and fiction. By using this framework, I hope to show that understanding models in terms of make-believe allows us to develop a coherent, general account of scientific modelling, one that explains what models are and how they represent the world.

The discussion will proceed as follows:

Chapter 1 will introduce the two main problems to be addressed. First, we will consider the ontological problems posed by theoretical modelling. As we shall see, some have tried to solve these problems by taking model-systems to be abstract entities. More recently, others have suggested that they are fictional entities, like unicorns or Count Dracula. I shall call these *indirect views* of modelling, since, on these accounts, scientists represent the world indirectly, via abstract or fictional entities. Although recent philosophy of science has seen a great deal of interest in the problem of scientific representation, there is also disagreement over precisely what the problem is. In fact, according to one influential view, 'there is no special problem of scientific representation' (Callender and Cohen, 2006, p. 67). In Chapter 1 I will argue that this view is mistaken, and show why the problem of scientific representation must still be faced.

In Chapter 2, I will introduce Walton's framework for understanding games of make-believe and show how it may be applied to scientific models. I will then go on to show how this *make-believe view* of modelling allows us to solve the ontological problems posed by theoretical models. As we will see, the make-believe view allows us to make sense of what scientists are doing when they model the world without positing any object that satisfies their modelling assumptions. On the make-believe view, there are no model-systems: scientists represent the world directly, not via abstract or fictional entities. We will also look more closely at how this *direct view* differs from those which compare model-systems to fictional entities.

Chapter 3 will focus on the problem of scientific representation. I will show that the make-believe view allows us to offer an account of representation which meets the requirements I introduce in Chapter 1. The account I suggest will point to a parallel between scientific models and works of fiction. There is now a growing body of work in philosophy of science that, in one way or another, seeks to compare models and fiction. Unsurprisingly, there are also those who have objected to this comparison. In Chapter 3, we will pause to consider these objections and see whether they present a problem for the make-believe view. We will then focus on a type of model that often causes difficulties for theories of scientific representation. These are models which are representational, but which represent no actual, concrete object. Unlike existing accounts, the make-believe view is able to make sense of these models, since it does not take scientific representation to be a relation.

As well as helping us to solve philosophical problems, any account of scientific models should also be able to provide a convincing account of the practice of modelling. Historians of science have offered us detailed studies of representation in scientific practice (e.g., Lynch and Woolgar, 1990), and of three-dimensional, physical models in particular (de Chadarevian and Hopwood, 2004). Physical models are often neglected by philosophers of science, however (exceptions include Sterrett, 2002 and Weisberg, forthcoming). Chapters 4 and 5 will focus on an important group of physical models in science, namely molecular models.

Chapter 4 examines the role of models in the work of the Dutch chemist J. H. van't Hoff (1852–1911). Winner of the first Nobel Prize in chemistry, van't Hoff was one of the founders of stereochemistry, the part of chemistry that concerns the spatial arrangement of atoms within molecules. The implications of van't Hoff's work were radical: at a time when even the existence of atoms remained controversial, van't Hoff's 'chemistry in space' claimed to show the way that atoms are arranged within molecules. And yet van't Hoff's work met with surprisingly little opposition. As we shall see, we may better understand the reception of van't Hoff's ideas by focusing on the cardboard models that he used to promote his work. I will argue that the make-believe view offers a framework with which to make sense of these early chemical models and the important role they played in the development of stereochemistry.

While historical studies are important, it is also helpful to be able to observe the way that models are used first-hand. Chapter 5 will assess the make-believe view through an empirical study of some molecular models in use today. I will suggest that the make-believe view gains support when we look at the way that these models are used and the attitude that users take towards them. Users' interaction with molecular models suggests that they imagine the models to be molecules, in much the same way that children imagine a doll to be a baby. Furthermore, I argue, users of molecular models imagine themselves viewing and manipulating molecules, just as children playing with a doll might imagine themselves looking at a baby or feeding it. Recognising this 'participation' in modelling, I suggest, helps us to understand the value of physical models and the bodily manipulation that they allow. It also points towards a new account of how models are used to learn about the world, through what I call *imagined experiments*.

By the end of the book, I hope that the comparison between models and children's dolls and toy trucks will not seem so strange after all. In fact, I believe that it will offer us a rather natural account of what scientific models are and how they are used. We sometimes describe modelling as a process of treating the world 'as if' it were a certain way, or say that it involves 'pretending' that a system obeys certain laws or assumptions. I hope to show how far this way of understanding modelling can get us.

1
Models and Representation

This chapter will introduce the two main problems addressed in this book. We will see that theoretical modelling presents us with a number of ontological puzzles (Section 1.2.1) and that many philosophers believe that solving these puzzles requires us to understand theoretical models as abstract or fictional entities (Sections 1.2.2 and 1.2.3). We will then look more closely at the problem of scientific representation (Sections 1.3.1–1.3.3). Recently, some have argued that this problem does not require a solution, since scientific models depend on other, more fundamental, forms of representation. At the end of the chapter, we will see what is wrong with this argument, and why the problem of scientific representation must still be faced (Sections 1.3.4 and 1.3.5).

First, however, we need to take a quick look at the different sorts of models that we find in science (Section 1.1).

1.1 Models

1.1.1 Physical models

When one thinks of models in the sciences, perhaps the first examples that come to mind are the 'ball-and-stick' models of molecules used in chemistry classes or the astronomical models of the solar system found in museum displays. I shall refer to such models as *physical models*. I use the term 'physical' only to indicate that they are actual, physical objects and not to suggest that they are constructed within the physical sciences. Early molecular biologists constructed models from metal and plastic representing the complex structures of protein molecules (de Chadarevian, 2002, 2004). Eighteenth century natural philosophers created intricate anatomical models of the body made from wax and wood (Mazzolini, 2004). Many physical models might also be referred to as 'scale models', like those used by engineers or architects to display or test a new design. But not all physical models are scale models. The famous Phillips machine represents the workings of the macro-economy by the ebb and flow of coloured water in a hydraulic system (Morgan and Boumans, 2004). And in Chapter 4, we will see that early chemical models were also not scale models.

1.1.2 Theoretical modelling

Most philosophical literature on modelling focuses not on physical models, but on *theoretical modelling*. In *Representing and Intervening*, Ian Hacking refers to the models of molecular biology, 'made with spring washers, magnets, lots of tin foil and such' and to the 'hold-in-your-hand models' of nineteenth century physics. But he goes on to say that '[m]ost generally [...] a model in physics is something you hold in your head rather than your hands' (1983, p. 216). Similarly, Ronald Giere expresses a commonly held view when he writes that

> the class of scientific models includes physical scale models and diagrammatic representations, but the models of most interest are *theoretical* models. (1999a, p. 5, emphasis in original)

Exactly what is meant by 'theoretical modelling' is often less clear. To understand the way in which I shall use the term, let us begin by considering a simple example. Suppose we are interested in predicting the motion of a bob bouncing on the end of a spring. Hooke's Law states that the force exerted by a spring is proportional to its extension. With this in mind, we might proceed to write down Newton's Second Law for a mass m, subject to a linear restoring force:

$$md^2x/dt^2 = -kx \qquad (1.1)$$

Here k is the 'spring constant' and x is the displacement from the 'equilibrium position'. Inputting the mass of the bob and the value of the spring constant, we could then solve our equation of motion to give the position of the bob at any given time after its release. We could also calculate that the period of oscillation of the bob, T, is given by the formula

$$T = 2\pi\sqrt{m/k} \tag{1.2}$$

And our predictions may well be rather accurate.

If pressed, however, we would readily admit that the equation we have written down is not, strictly speaking, true of the bouncing bob. By writing the Second Law in the way that we did, we assumed, for instance, that no air resistance acts on the bob. We are aware that our assumption is false: the bob is not bouncing in a vacuum. We also assumed that the gravitational field acting on the bob is uniform, whereas, in fact, it increases as the bob moves closer to the earth. Indeed, we make many such assumptions when we model the bouncing spring: we take the bob to be a point mass m subject only to a uniform gravitational field and a linear restoring force exerted by a massless, frictionless spring with spring constant k attached to a rigid surface. By describing the system in this way, we are able to apply our equation of motion and calculate predictions for the bob's behaviour. But we are aware that this description is false, and that a full description of the system would make reference to the air resistance on the bob, the mass of the spring, and so on.

This is an instance of theoretical modelling. In order to calculate predictions for the behaviour of the bouncing spring we *model* the spring as a simple harmonic oscillator. Of course, we do not construct an actual, physical model, as an engineer might construct a scale model of a bridge. Instead, we simplify or idealise the bouncing spring so that we can apply the theoretical laws available to us. This is often expressed by saying that we treat the spring 'as if' it were a point mass m, subject to a uniform gravitational field, and so on. Of course, other more complex theoretical treatments are available to us in this case. For example, we could try to account for air resistance by taking the bob to be subject to a further force proportional to its velocity. But even these more complex treatments invoke some false assumptions. And our initial model may be perfectly adequate for the task in hand.

Theoretical modelling is extremely common in the sciences. As Lawrence Sklar puts it,

> [w]e continually see scientists describing systems 'as if' they were systems of some more familiar kind [...]. Treating the systems

in this 'pretend' way often provides a royal route into gaining predictive and explanatory control over the systems in question. (2000, p. 71)

Familiar examples include the billiard ball model of gases, Bohr's model of the atom, the liquid-drop model of the nucleus, the Lotka-Volterra model of predator-prey interaction, the London model of superconductivity, the MIT-Bag model of quark confinement, general equilibrium models of markets and Newtonian models of the solar system. In each of these cases, the term 'model' is used by the scientists themselves. However, I shall use 'theoretical modelling' in a broader sense, to include any case in which scientists deliberately simplify or idealise a system in order to explain or predict its behaviour, whether or not the outcome of this process is commonly referred to as a model. Each of the familiar examples above counts as an instance of theoretical modelling in this sense: the billiard ball model of a gas, for example, assumes that the molecules of the gas are not acted upon by gravity and that they do not act upon each other between collisions, while Newtonian models of the solar system typically take the planets to be perfect spheres with even mass distribution and assume that they are subject only to the sun's gravitational field. But the activity of theoretical modelling extends far beyond such well-known examples. Indeed, it is almost always the case that a system must be simplified or idealised in some way before it can be given a theoretical treatment.

This point is put perhaps most forcefully by Nancy Cartwright. In *How the Laws of Physics Lie* (1983), Cartwright argues against what she calls 'covering-law models of explanation'. According to these covering-law models, 'we fit a phenomenon into a theory by showing how various phenomenological laws which are true of it derive from the theory's basic laws and equations' (1983, p. 16). Cartwright points out that in most cases, such explanations are simply unavailable. She offers the example of quantum mechanics. In order to explain the behaviour of a system using quantum mechanics one must formulate a Hamiltonian for that system (a system's Hamiltonian is an expression of its total energy). However, the theory provides Hamiltonian functions only for what Cartwright calls 'highly fictionalised objects', like a 'free particle in a box', square potential well or linear harmonic oscillator (ibid., p. 136). The strict deductive relations demanded by covering-law accounts are available only for these highly idealised systems. How are we to explain the behaviour of real systems?

In place of the covering-law view of explanation, Cartwright offers her own 'simulacrum account'. On this account, in order to bring a real world situation under a fundamental law or equation, we first 'start with an *unprepared* description which gives as accurate a report as possible of the situation' (ibid., p. 15, emphasis in original). We then 'convert this into a *prepared* description' (ibid., emphasis in original), that is, one which can be dealt with by the theory. For example, we might describe a hydrogen atom as an electron and a proton subject to a Coulomb potential, thus neglecting the spin of the electron, the electromagnetic field, any effects on the atom from its environment, and so on. Although, 'ideally the prepared description should be true to the unprepared' (ibid.), in general it will not be: 'our prepared descriptions lie' (ibid., p. 139). This process of '"preparing a description" is exactly what we do when we produce a model for a phenomenon' (ibid., p. 15). And according to Cartwright, 'a model – a specially prepared, usually fictional description of a system under study – is employed whenever a mathematical theory is applied to reality' (ibid., p. 158).

In some cases, like that of the bouncing spring, the account that scientists offer of the system being modelled is derived from some existing theory, such as Newtonian mechanics. The term 'theoretical modelling' is sometimes reserved only for such cases. I shall not restrict the term in this way. Recent case studies have revealed that scientists must often go beyond existing theory in order to model a system. Indeed, Mary Morgan and Margaret Morrison claim that 'models occupy an autonomous role in scientific work' (1999, p. 10). Similarly, Cartwright now argues against what she calls 'the vending machine view', on which models are straightforwardly dispensed from theory (1999, p. 184). I will use the term 'theoretical' only to indicate that the scientists do not construct an actual, physical object that serves as their model, not to imply that the model is derived from some existing theory.

1.2 The ontology of theoretical modelling

1.2.1 The problem: missing systems

In discussing theoretical modelling, I have sometimes talked about theoretical *models*. I have not said what these are, however. This omission is common in discussions of theoretical modelling. We have no problem explaining what we mean when we talk about a physical model: we can simply point to the lump of wood or plastic in front of us. What do we mean when we talk about theoretical models?

Unlike the engineer who constructs a scale model of a bridge, we do not build any physical object that satisfies our prepared description or equation of motion. Indeed, no actual, concrete object satisfies our assumptions: no real bob is a point mass or is subject to a perfectly uniform gravitational field, no real spring has zero mass or is entirely free from friction, and so on. And yet, we often think and talk about theoretical and physical modelling in similar ways. The engineer constructs a scale model of a bridge in the hope that, by learning about the properties of the model, she will learn something about the bridge itself. We often discuss theoretical modelling in the same way. What we do when we model the bouncing spring, we might say, is to construct a simplified and idealised version of it. We then investigate the behaviour of this idealised bouncing spring in order to learn about the more complex behaviour of the spring itself.

Cases of theoretical modelling thus present a number of problems. One problem concerns the various prepared descriptions and theoretical laws that we write down when we model a system. For example, consider the prepared description that we formulate when we model the bouncing spring: it takes the bob to be a point mass and the spring to exert a linear restoring force. It seems we cannot regard this as a straightforward description of the bouncing spring system: we do not claim that the bob is a point mass, for example, or that the force the spring exerts is linear. We make neither of these claims because we know them both to be false. The same is true of our equation of motion (1.1). Sometimes, idealising assumptions such as these are referred to as 'misdescriptions'. For example, Ronald Laymon calls the process of idealisation 'the misdescription of things' (1998). When we refer to something as a misdescription, however, we usually mean simply that it is a description that is incorrect or inaccurate in some way; it is a description that turned out to be false. Our 'prepared description' does not seem to be an attempt to describe the system, either a successful or an unsuccessful one.

Our prepared description and equation of motion are an example of what Martin Thomson-Jones (2007, 2010) calls a *description of a missing system*: they look just like a description of some actual, concrete object, and yet we realise that there are no such systems that would satisfy this description. Furthermore, much of our subsequent talk about theoretical modelling seems to assume that there is such an object. Thus, even though there is no actual, concrete, simple harmonic oscillator that satisfies our prepared description and equation of motion, we often make statements that appear to refer to such an object. For example, we might remark that 'the bob oscillates sinusoidally' or 'the force exerted by the spring is linear'. Thomson-Jones calls the practice of talking as if there were objects that satisfy descriptions of missing systems *the face-value practice*.

The face-value practice is common in discussions of theoretical modelling. For example, a predator-prey model might be said to contain only two species, predator and prey, and we might be told that the prey reproduce exponentially if they are not subject to predation. We may be told that the electrons in the Bohr model occupy fixed orbits around the nucleus, or that the collisions in the billiard ball model are perfectly elastic. Scientists say that the objects in a model are isolated from the environment, or that they lose energy over time. As R. I. G. Hughes notes, 'we talk about the behaviour of the model, but rarely about the behaviour of a picture or map' (1999, p. 126). Cartwright writes that 'fundamental laws do not govern objects in reality; they govern only objects in models' (1983, p. 18). 'The fundamental laws of the theory', she says, 'are true only of the objects in the model, and they are used to derive a specific account of how these objects behave' (ibid., p. 17). Since she stresses that there is no actual, concrete object that satisfies the fundamental laws of the theory, it seems that in talking about 'objects in the model', Cartwright is engaging in the face-value practice. Similarly, Margaret Morrison writes that

> when we want to gain information about the behaviour of a real pendulum we can't simply apply laws of Newtonian mechanics directly; rather we need a way of describing an idealised pendulum – a model – and then applying the laws to the model. (1999, p. 48)

The face-value practice is thus reflected in an ambiguity in the way that we use the term 'model'. Often, when we talk about a model, we simply mean the various assumptions and equations that the scientist writes down. In this sense, our model of the bouncing spring is just our prepared description and equation of motion. Sometimes, however, we use 'model' in a different sense. In this second sense, when we talk about a model, we mean to indicate not any set of assumptions or equations, but instead some sort of idealised or simplified system that satisfies those assumptions and equations. By using 'model' in this way, it seems, we engage in the face-value practice. In order to avoid confusion, I will resist using 'model' in this second sense, but will instead talk of the 'model-system'. Thus, while our model of the bouncing spring is our description and equation, these appear to describe a model-system consisting of a point mass, massless spring, and so on.

With this in mind, we can now set out three questions that any account of theoretical modelling must address. The first asks how we are to interpret the assumptions and equations that scientists write down when they model a system: how should we understand the descriptions of missing systems that appear in theoretical modelling? The second question concerns the face-value practice: how should we interpret statements scientists subsequently make while modelling a system, when they seem to assume that there is a model-system that satisfies their assumptions? We may also pose a third question. As well as talking as if there were a model-system, scientists often speak as if they may *learn* about that system. Consider again the bouncing spring model. Just as the engineer may learn that her scale model withstands a certain load, so it seems that we may discover that the bob in our model-system oscillates sinusoidally or that its period of oscillation is given by the formula (1.2). A third question for our account of theoretical modelling asks us to make sense of this: how are we to explain how we can learn about a model-system?

In sum: given that model-systems are not actual, concrete parts of the world, how is it that scientists seem to be able to talk about them and learn about their properties?

1.2.2 Description-fitting objects: abstract objects

One way to answer these questions is to insist that while no actual, concrete object satisfies our prepared description and equation of motion, there is some other object that does satisfy them. To borrow another term from Thomson-Jones, we might posit *description-fitting objects* to serve as model-systems.

This is the route taken by Ronald Giere (e.g., 1988, 1999a, 2004). Like Cartwright, Giere observes that most theoretical laws or 'principles', such as Newton's laws of motion or the principles of quantum mechanics, are true only of highly idealised systems. But he goes on to draw a different lesson for the interpretation of these principles. As Giere puts it,

> I disagree only with Cartwright's general description of her conclusion, not its real content. In my view the general laws of physics, such as Newton's laws of motion and the Schrödinger equation, cannot tell lies about the world because they are not really statements about the world. They are, as Cartwright herself sometimes suggests, part of the characterisation of theoretical models, which in turn may represent various real systems. (1988, p. 90)

Consider a theoretical principle such as $f = ma$. If we wish to use this principle to predict the behaviour of a system we must fill in the force function, just as we must select a Hamiltonian before we apply the Schrödinger equation. Again, classical mechanics provides force functions only for highly idealised systems, and explaining the behaviour of a real system, such as the bouncing spring, involves making a number of false assumptions about that system. In light of this, Giere asks how we should interpret principles such as $f = ma$. We appear to run into difficulty if we regard this as an empirical statement. For once we fill it in with a specific force function, we almost always arrive at an equation that is known to be false of the system we are interested in. For Giere, the lesson is clear:

> If we insist on regarding principles as genuine statements, we have to find something that they describe, something to which they refer. The best candidate I know for this role would be a highly abstract object, an object that, by definition, exhibits all and only the characteristics specified in the principles. (2004, p. 745)

According to Giere, then, theoretical principles like $f = ma$ are not statements about the world, but definitions of abstract objects. When we wish to apply these principles to the world, we must add various specific conditions. For example, when we model the bouncing spring we fill in the Second Law with the specific force function

$f = -kx$ to give our equation of motion (1.1). We also identify x as the displacement of a mass on a spring, rather than, say, the displacement of a pendulum or the voltage in an electrical circuit. Taken together, according to Giere, theoretical principles and specific conditions define more specific abstract objects. These objects are theoretical models (or, in my terminology, model-systems).[1]

Giere's account thus offers one answer to the problems posed by theoretical modelling. It does so by claiming that, while there is no actual, *concrete* object that satisfies our prepared description and equation of motion, there is an actual, *abstract* object that does. It is this object that is the model-system. This account offers us a way to make sense of our prepared description and equation of motion: they may be understood as *definitions* of a nonlinguistic abstract object. It also appears to make sense of our commonsense talk about theoretical modelling, by providing us with an analogue of the engineer's scale model: a nonlinguistic abstract object. When scientists engage in the face-value practice, we might interpret their utterances as claims about this abstract object. Finally, Giere's account might also provide us with a way of explaining how we can learn about a theoretical model: just as the engineer investigates the properties of her scale model, so we might investigate the properties of our abstract simple harmonic oscillator.

Giere's account has seemed attractive to many. It is not without problems, however. For example, Thomson-Jones (2007, 2010) argues that, since there are no actual, concrete objects that fit our prepared description and equation of motion, the abstract objects posited by Giere's account cannot be spatiotemporal entities. But then how can such objects possess the spatiotemporal properties we appear to attribute to them, such as oscillating with period T?[2] Partly in response to such concerns, some have suggested that, rather than abstract objects, model-systems should be understood in the same way as fictional characters, like Emma Bovary or Count Dracula. Let us now turn to consider this view.

1.2.3 Description-fitting objects: fictional characters

Consider the following passage, from the novel *Dracula*:

> The Count smiled, and as his lips ran back over his gums, the long, sharp, canine teeth showed out strangely. (Stoker 1897/1994, p. 33)

Thankfully, there is no actual, concrete object that this passage represents: Count Dracula, as we might put it, is 'only a fictional character'. And yet, this passage, along with the rest of the novel, still seems to represent Count Dracula. Intuitively, we might say that the novel is 'about' Count Dracula. While reading the story, we seem to imagine things about the Count, such as that he sucks blood and has long, sharp teeth. If we say 'Dracula has long teeth', it appears we assert something true about him, while if we say 'Dracula is vegetarian', it seems that we assert something false. How can we make sense of all this if there is no Count Dracula?

Works of fiction that involve fictional characters like Count Dracula therefore present us with a number of philosophical puzzles. A variety of different solutions has been proposed. These fall into roughly two camps: *realist* and *antirealist*. *Realists* argue that, even if he is not a flesh-and-blood living (or rather, undead) thing, we must grant that Count Dracula does exist in *some* sense. Count Dracula, along with Hamlet, Madame Bovary, Middle Earth and the rest are therefore given a place in our ontology as *fictional entities*. What exactly are fictional entities? Here again a number of different accounts have been proposed. Alexius Meinong (1904/1960) famously distinguishes *being* from *existence*. On his account, Count Dracula is an object that possesses all the properties that we normally take him to have, such as being a vampire and sucking blood; he simply lacks the property of existence. Other realists, such as Peter van Inwagen (1977), take fictional entities to be abstract, rather than nonexistent, entities.

Antirealists aim to show how we can understand works that invoke fictional characters, and our talk about them, without granting the existence of fictional entities. The best known antirealist theory follows Bertrand Russell (1905/1956) and analyses an utterance like 'Dracula sucks blood' as something like the claim that 'there exists exactly one x such that x satisfies the Dracula-description and x sucks blood'. The statement 'Dracula sucks blood' is then perfectly meaningful and contains no reference to any fictional entity. This theory is not without problems, however. For although the sentence 'Dracula sucks blood' is meaningful in his analysis, it is also false, and so on a par with any number of other claims about Dracula, like 'Dracula is vegetarian'. And yet, intuitively there seems to be a considerable difference between these claims.

Fictional entities, then, are objects proposed by realists in order to make sense of works involving fictional characters. In order to avoid confusion, I shall use the expression *fictional characters* in an ontologically neutral sense to refer to the characters, places and objects apparently represented in works of fiction, reserving the term *fictional entities* for the various different entities posited by realists.[3] Thus, while everyone agrees that works of fiction involve fictional characters, they differ on how we should make sense of this: realists believe we must posit some kind of object that those works represent and which we talk about (fictional entities), while antirealists do not.

Within recent work on models, a number of authors have been struck by the apparent parallels between theoretical modelling and works of fiction that involve fictional characters (e.g., Contessa, 2010; Frigg 2010a, 2010b; Godfrey-Smith, 2006). Like scientists' prepared descriptions, it seems the passage quoted above from *Dracula* is also a description of a missing system: there is no actual, concrete object that it describes. Moreover, just as scientists often talk as if there were an object that satisfied their prepared description, so we also talk as if there were a Count Dracula: we say that Dracula has long teeth and that he sucks blood. We saw above that some have criticised Giere's view on the grounds that we often ascribe spatiotemporal properties to model-systems. We certainly have no problem attributing spatiotemporal properties to fictional characters: we say that Count Dracula is tall, for example, or that Sherlock Holmes lived in 221B Baker Street.

These observations motivate what I will call the *indirect fictions view* of modelling. (The reasons for this label will become clearer shortly.) According to proponents of this view, scientists' prepared descriptions and equations of motion give rise to model-systems, and these model-systems are to be understood in the same way as fictional characters like Count Dracula. Of course, the biggest challenge for these accounts concerns the ontology of model-systems. As we have seen, fictional characters present us with difficult philosophical puzzles, and there are different ways of understanding them. Some think that there are such things as fictional entities (and offer very different accounts of what these entities are), while some do not. In itself, then, simply comparing model-systems with fictional characters does not seem to get us very far.

Proponents of the indirect fictions view respond to this problem in a number of ways. Some look to existing theories of fictional characters. Thus, Martin Thomson-Jones (2007) explores the prospects for applying a number of different accounts of fiction to models (although he does not endorse any of them), while Roman Frigg (2010a, 2010b) draws on Walton's theory. Instead of relying on existing theories of fiction, Gabriele Contessa (2010) develops his own account of fictional entities. Peter Godfrey-Smith (2006) adopts a different strategy. Rather than defending any particular account of the ontology of fictional characters, he suggests that we might remain content with accepting such objects as part of the 'folk ontology' of scientific modelling, even if in the long run we require an account of these objects 'for general philosophical reasons' (2006, p. 735). Ronald Giere also endorses this strategy. Although originally a proponent of the view that model-systems are abstract objects, Giere (2009) has recently suggested that he too is willing to think of them as akin to fictional characters. Like Godfrey-Smith, however, Giere argues that philosophers of science need not be too concerned with the question of exactly what such entities are.

We shall consider these accounts in detail in Section 2.4. First, we must turn to examine the other key problem facing theories of scientific modelling.

1.3 The problem of scientific representation

1.3.1 The problem

Consider the physical models mentioned in Section 1.1. One feature that all of these models share is that they are used to *represent* some object or event in the world. The engineer's scale model represents the bridge. Crick and Watson's model represents the DNA molecule. The Phillips model represents the economy. Similarly, it seems that we often represent the world through theoretical modelling. Consider our model of the bouncing spring. We have already seen that we do not describe the spring when we model it. Nevertheless, it seems that we do represent the spring in some sense. Intuitively, we might say that we represent it *as* a simple harmonic oscillator. The problem of representation for scientific models asks how such cases of representation work.

When we turn to theoretical modelling, the problem of representation takes different forms, depending on our view of the ontology of modelling. For example, suppose we were to adopt Giere's original view. In this view, scientists' prepared descriptions and theoretical laws do not represent the system being modelled directly. Instead, they define a nonlinguistic abstract object (the model-system). Nevertheless, Giere still thinks that scientists represent the world when they model it. His account therefore makes representation in theoretical modelling a two-stage process. First, prepared descriptions and theoretical laws define abstract objects. Second, these objects represent the system being modelled. If we adopt this view, then, understanding representation in theoretical modelling requires us to understand the nonlinguistic representation relation that is purported to hold between abstract objects and the systems they model. I shall refer to such accounts as *indirect* views of theoretical modelling (see Figure 1.1).

Those who take model-systems to be like fictional characters also hold a version of the indirect view. According to this view, scientists' prepared descriptions and theoretical laws give rise to model-systems, just as works of fiction give rise to fictional characters. And yet, like Giere, proponents of this view do not think that science is only in the business of representing fictional model-systems; they grant that scientists sometimes represent the world in theoretical modelling. In their view, however, when scientists do represent the world, they do so indirectly, via the model-system. Like Giere, then, proponents of this view adopt an indirect view of theoretical modelling: prepared descriptions and equations of motion give rise to model-systems and these model-systems, in turn, represent the world. Different views have been advanced regarding exactly how model-systems represent the world. Godfrey-Smith (2006), for example, follows Giere in talking of resemblance between model-systems and the world, while

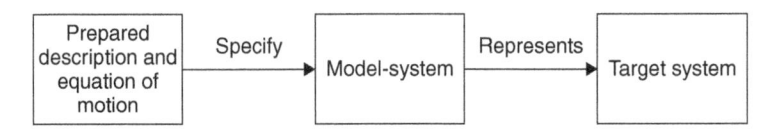

| Prepared description and equation of motion | —Specify→ | Model-system | —Represents→ | Target system |

Figure 1.1 Indirect views of theoretical modelling

Frigg (2010b) speaks of a 'key' which specifies how facts about the model-system are translated into claims about the real system.

In the course of addressing the problem of representation for scientific models, it will often be helpful to look to another representational device, namely pictures. Like models, many pictures are representational, and some represent actual objects or events. Jacques-Louis David's *Napoleon Crossing the Saint Bernard* represents Napoleon. Constable's *Salisbury Cathedral from the Meadows* represents Salisbury Cathedral. The problem for theories of pictorial representation is to understand how they do this. In itself, Constable's painting is merely a set of brushstrokes on a piece of canvas. And yet, it depicts horses pulling a cart through a stream, and in the distance, the cathedral itself beneath a rainbow. How does the painting achieve this? In virtue of what does Constable's painting represent the cathedral? The problem of representation for scientific models may be presented in the same way. The reconstruction of Crick and Watson's original DNA model in the Science Museum is simply a collection of metal rods and plates held in place by clamps. And yet it represents the complex helical structure of the DNA molecule. How does it do this? In virtue of what does the model represent the molecule?

We have a name for the sort of representation pictures provide. We say that David's painting *pictures* or *depicts* Napoleon, and that Constable's landscape *depicts* Salisbury Cathedral. Of course, pictures represent in other ways too, apart from depiction. David's painting might be said to represent the glory of France, or Constable's 'the culmination of his numerous treatments of Salisbury Cathedral' (National Gallery, 2008). Such is the looseness of the term 'represent'. But it is one particular form of representation that pictures offer, namely depiction, which theories of pictorial representation seek to explain. We lack a name for the way that models represent. If we say merely that models represent their objects, then we are likely to be misled, for the word 'representation' is used in so many different ways. Crick and Watson's model might also be said to represent the greatest achievement of British science or Bohr's model a belief in the simplicity of the atomic realm. In analogy to pictorial representation, then, we might label the form of representation we are interested in *model-representation*. Crick and Watson's model, we shall say, *model-represents* the DNA molecule and Bohr's model *model-represents* the atom.

We must be careful here, however. The variety of scientific models is remarkable. What reason do we have for thinking that all of these models represent in the same way?[4] Does Crick and Watson's model represent the DNA molecule in the same way as Bohr's model represents the atom, for example, or our model above represents the bouncing spring? Might there not be many forms of model-representation? Here the contrast with depiction is telling. The variety of things we call 'pictures' is also remarkable. It includes figurative paintings, Impressionist landscapes, political cartoons, children's drawings, stick figures and much more. And yet, despite their obvious differences, it is often thought that there is one form of representation that is common to all of these pictures, namely depiction. Intuitively, many people feel that whatever the obvious differences between, say, a landscape by Constable and a frame from a *Tom and Jerry* cartoon, both are depictions. It is this form of representation that theories of depiction seek to analyse. The sheer variety of pictures is partly what makes this analysis so difficult. I do not think that we have the same intuition for scientific models. Whether or not there is a form of representation common to both Crick and Watson's and Bohr's models, for example, would seem to be an open question that a theory of scientific representation must address.

We must not assume, then, that there is one form of representation that is common to all scientific models: there may turn out to be many different forms of model-representation. And we should also be careful not to assume that any of these forms of representation are unique to scientific models. Any of the forms of model-representation that we identify, or even all of them, may turn out to be a species of some broader form of representation, used either within or outside of science. Scientists employ many forms of representation also found outside of the sciences, such as images, maps and diagrams. Similarly, perhaps we shall find that scientific models represent in the same way as other representational entities used elsewhere, such as pictures or sculptures. In this case, model-representation might turn out to be a form of pictorial representation.

Our account of model-representation need not, therefore, answer the question of what makes a model *scientific*. This may be an interesting question in its own right but it is one which a theory of representation for scientific models does not have to answer. We want to know what form or forms of representation are employed in scientific modelling. As we have observed, it may turn out that some of these forms of representation are also employed by other representational devices outside of science, in art or in everyday life. But our theory of representation does not need to go on to say how, if at all, scientific models differ from these other representational devices. A similar situation exists with regard to pictures. Accounts of depiction aim to provide an analysis of all pictures, including stick figures and children's drawings as well as Picassos and Rembrandts. These accounts do not need to tell us what, if anything, makes some of these pictures art while others are merely doodles.

Often, the task of providing an account of how models represent is taken to be that of analysing a *relation* between a model and the part of the world that it represents. The task for theories of depiction is often presented in the same way. A theory of depiction, it is said, must tell us what the relation is between a picture and its subject, in virtue of which it depicts that subject. The difficulty with presenting the task in this way, of course, is that many pictures have no actual subject. And yet it seems that a picture of a unicorn is still depictive, even though there is no unicorn that it depicts.

The same problem arises for scientific models.[5] Hughes writes that

> the characteristic – perhaps the only characteristic – that all theoretical models have in common is that they provide representations of parts of the world, or the world as they describe it. (1997, p. S325)

In fact, however, there are many models which represent no actual object. Obvious examples are models of discredited entities, like mechanical models of the ether. But, as we shall see in Chapter 3, there are many other sorts of cases. How are we to deal with such models? One option open to us is simply to deny that they are representational. This seems unsatisfactory: intuitively, I think, we regard ether models as representational, just as we regard pictures of unicorns as representational. But if we want to allow that such models are representational, then we are faced with a dilemma: either we postulate some entity that they represent or we cease to think of model-representation as essentially relational. I shall consider this problem in depth in Chapter 3. For the moment, we may simply note that we should not assume that representation is a relation if we do not wish to commit ourselves to either way out of this dilemma.

To sum up: many scientific models are representational. Some represent actual objects or events. The problem of scientific representation asks how they do this. Why does Crick and Watson's model represent the DNA molecule, or our model represent the bouncing spring? There may turn out to be many different forms of model-representation. Any, or even all, of these forms of representation may be employed by other representational devices, apart from scientific models. We want an account of each of these forms of model-representation. Theories of depiction aim to state conditions that are necessary and sufficient for something to be a depiction. Similarly, if possible, we want to provide a set of conditions that are both individually necessary and jointly sufficient to establish an instance of each form of model-representation that we identify.

1.3.2 Misrepresentation

It is important to distinguish the problem of representation for scientific models from a closely related question. That is the question of what makes a model accurate (or, perhaps, correct or realistic). Most models are inaccurate (or incorrect or unrealistic) in some way. Often this is deliberate. When we model the bouncing spring, we neglect the effects of air resistance. Sometimes, inaccuracy is unintentional: before building their famous DNA model, for example, Crick and Watson constructed and rejected a number of different models. For our present purposes, the important point to notice is that inaccurate models are still representations, and so they must be accommodated by our theory of model-representation. Both our simple model and a model that accounts for air resistance represent the spring, and Crick and Watson's early efforts, like their final double-helical model, all represent the DNA molecule.

Many people share the intuition that an account of scientific representation should accommodate inaccurate, as well as accurate, models. For example, Mauricio Suárez writes that

> a good theory [of scientific representation] may provide us with insight into some of the features that are normally associated with scientific representations such as accuracy, reliability, truth, empirical adequacy, explanatory power; but [...] we shall not assume that

> this is a requirement. In other words, we shall not require a theory of representation to mark or explain the distinction between accurate and inaccurate representation, or between a reliable and unreliable one, but merely between something that is a representation and something that is not. (2003, p. 226)

Similarly, according to Frigg, it is an important requirement for any theory of representation that it allow for

> the possibility of misrepresentation. A tenable theory of scientific representation has to be able to explain how misrepresentation is possible. Misrepresentation is common in science. Some cases of misrepresentation are, for all we know, plain mistakes (e.g., ether models). But not all misrepresentations involve error. Many models are based on idealising assumptions of which we know that they are false. Nevertheless these models are representations. A theory that makes the phenomenon of misrepresentation mysterious or impossible must be inadequate. (2006, p. 51)

We should be careful to distinguish between two types of misrepresentation that are alluded to in this passage. The principal failing of ether models, it seems, is that they fail to represent any actual object. I have already noted the importance of including such models in our account, and we shall return to consider them in more detail in Chapter 3. At present, we are concerned with a different type of misrepresentation. This is the misrepresentation that occurs with models that *do* represent an actual object, but which do so in a way that is erroneous or inaccurate. Like Frigg and Suárez, I claim that such models are still representations.

Some will disagree with this characterisation of the problem facing accounts of scientific representation. As we have already observed, use of terms such as 'representation' or 'depiction' is often vague and subject to dispute. (Are doodles really depictions? What about stick men?) For some, 'representation' carries implications of realism, or at least empirical adequacy, when used with regard to scientific models. Although I do not understand the term in this way, I do not, of course, deny that the question of what makes one model more accurate than another is an important one. I claim only that this question need not be addressed by our theory of representation for models, which is concerned with the prior question of what makes something a model-representation. Once we understand how models represent, we will want to make further distinctions among models, distinguishing good from bad along various different dimensions. If someone wishes to reserve the term 'representation' for those models that fall only on the good side of one or more of these divides, then we needn't quibble too much. The more important point is that our account of representation should provide us with the resources to make these distinctions amongst models.

The situation is similar with pictures. We often judge the realism of pictures, counting a Rembrandt more realistic than a cave painting or a Picasso.[6] This raises the question of what makes one picture more realistic than another. Many theories of depiction suggest a natural answer to this question. For example, the view that pictures depict in virtue of resemblance suggests an account of realism: the more a picture resembles its object, the more realistic it is. However, the question of what constitutes realism need not be addressed by a theory of depiction. Unrealistic pictures are still pictures; they still depict their objects. It is this which a theory of depiction must try to understand.

1.3.3 Two kinds of solution

It will be helpful to distinguish two different forms that our account of representation in modelling might take. I shall refer to these different types of accounts as *derivative* and *nonderivative*. A *derivative account* of model-representation would attempt to explain how models represent in terms of some other form of representation, such as mental or linguistic representation. It would attempt to show how the representational power of models derives from some other form of representation. By contrast, a *nonderivative account* would attempt to explain how models represent in nonrepresentational terms.

Some accounts of depiction will help to make this distinction clearer. Consider the most straightforward *resemblance* theory of depiction, attacked by Nelson Goodman (1976). This account claims that *A* is a depiction of *B* if and only if *A* is appreciably similar to *B*. This is a nonderivative account of depiction: it attempts to explain depiction in terms of those (nonrepresentational) properties that are shared by a picture and its object. By contrast, consider the reconstruction of Plato's account of depiction offered by Alan Goldman. According to this view,

> a picture represents an object if and only if (a) its artist successfully intends by marking a surface to create a visual experience that resembles that of the object, (b) such that the intention can be recovered from the experience, perhaps together with certain supplementary information, and (c) the object can be seen in the picture. (Goldman, 2003, p. 194)

This account attempts to explain depiction in terms of, amongst other things, the intention of the artist to create a certain visual experience of an object. It does not provide an account of mental representation that can explain what it is for the artist to have this intention. (And, of course, depending on one's theory of visual perception, one may think that the account invokes other representational mental entities.) Nevertheless, if this account is successful, it will have reduced the problem of depiction to some other problem (or problems) concerning mental representation. To take another example, consider Kendall Walton's *make-believe theory* of depiction (1990, Chapter 8). I will be discussing Walton's theory of depiction in Chapter 5. For now, all that is important is that for Walton, pictures are used in a certain sort of game of make-believe, in which the viewer imagines of her act of looking at the picture that it is an instance of looking at the object. Again, this is a derivative account: it aims to explain depiction in terms of the representational capacities of mental states (in this case, imagination).

1.3.4 Does the problem exist?

The problem of scientific representation is now the focus of a burgeoning literature in the philosophy of science (e.g., Bailer-Jones, 2003, 2009; Chakravartty, 2010; Contessa, 2007; Downes, 2009; Elgin, 2009, 2010; French, 2003; Frigg, 2006; Giere, 1999b, 2004, 2010; Hughes, 1997; Knuuttila, 2005; Suárez, 1999, 2003, 2004; van Fraassen, 2008; Weisberg, 2007). However, in a recent paper, Craig Callender and Jonathan Cohen (2006) have argued that this attention is unwarranted. In fact, the title of their paper declares, 'there is no special problem about scientific representation' (2006, p. 67). Clearly, if Callender and Cohen are right, then much of the discussion in this book is redundant. In this section, however, I shall attempt to show why they are wrong. In doing so, I also hope to clarify further the nature of the problem that faces us.

Callender and Cohen argue that we should approach the problem of scientific representation from a stance which they label *General Griceanism*. According to this view,

> among the many sorts of representational entities (cars, cakes, equations, etc.), the representational status of most of them is derivative from the representational status of a privileged core of representations. The advertised benefit of this General Gricean approach to representation is that we won't need separate theories to account for artistic, linguistic, ... and culinary representation; instead, the General Gricean proposes that all these types of representation can be explained (in a unified way) as deriving from some more fundamental sorts of representations, which are typically taken to be mental states. (ibid., p. 70)

A General Gricean account of representation therefore consists of two stages:

> First, it explains the representational powers of derivative representations in terms of those of fundamental representations; second, it offers some other story to explain representation for the fundamental bearers of content. (ibid., p. 71)

As an example of how the General Gricean programme may be implemented, Callender and Cohen outline what they call *Specific Griceanism*. In particular, they present a Specific Gricean account of linguistic representation. The account attempts to explain what it is for a speaker S to mean something by making a certain utterance U in terms of the speaker's intention to bring about a certain belief in the person H who hears the utterance:

> In uttering U, S means that p iff [if and only if], for some H, S utters U intending in way...to activate in H the belief that p. (ibid., p. 72)

This account thus aims to 'reduce the notion of speaker meaning for linguistic tokens to specific mental states of producers/hearers of these tokens' (ibid.). If successful, the account would complete the first stage in a Specific Gricean explanation of linguistic representation. The second stage would be to offer an account of mental representation.

Callender and Cohen are not committed to Specific Griceanism, which they recognise to be controversial. However, they are committed to General Griceanism, and in particular, to a General Gricean account of representation for scientific models. And it is by adopting this General Gricean position that Callender and Cohen believe we may 'solve or dissolve the so-called "problem of scientific representation"' (ibid., p. 67):

> Our proposal [...] is that scientific representation is just another species of derivative representation to which the General Gricean account is straightforwardly applicable. This means that, while there may be outstanding issues about *representation*, there is no special problem about *scientific* representation. (ibid., p. 77, emphasis in original)

Callender and Cohen offer little argument in support of General Griceanism, or its application to representation by scientific models. However, let us for the moment suppose that we were to accept their proposal. What would this mean for our enquiry into the way that models represent? Given what we have been told of the General Gricean approach, we might expect the first stage of this enquiry to be that of providing an account of how models represent in terms of some other, more fundamental form of representation. In the terminology I introduced in Section 1.3.3, we would expect to be asked to provide a derivative account of model-representation. According to Callender and Cohen, providing this derivative account 'amounts to a relatively trivial trade of one philosophical problem for another' (ibid., p. 73). And yet, if we could take this first step then we would have at least reduced the problem of explaining how models represent to the problem of explaining some other form of representation. And we might even feel at this stage that our work as philosophers of science was complete. The second step, of providing an account of the more fundamental form of representation, might be left to those working in the philosophy of mind or language. Furthermore, as

Callender and Cohen themselves admit, the Specific Gricean account of linguistic representation has hardly proved uncontroversial.

Immediately after they propose that we adopt a General Gricean approach to explain how models represent, however, Callender and Cohen expand on their claim in the following way:

> In particular, we propose that the varied representational vehicles used in scientific settings (models, equations, toothpick constructions, drawings, etc.) represent their targets (the behavior of ideal gases, quantum state evolutions, bridges) by virtue of the mental states of their makers/users. For example, the drawing represents the bridge because the maker of the drawing stipulates that it does, and intends to activate in his audience (consumers of the representational vehicle, including possibly himself) the belief that it does. (ibid., p. 75)

This further claim comes as a surprise. Rather than being asked to take the first step in our General Gricean account of scientific representation we are told that this step has already been taken. We do not need to provide a derivative account of representation for models. In fact, this account has a very simple form: all that is required for a model to represent its target is that the user of the model stipulates that it does, and that he intends to bring about the belief that it does. Consequently, Callender and Cohen claim, 'scientific representation [...] is constituted in terms of a stipulation, together with an underlying theory of representation for mental states' (ibid., p. 78). The representational relation between a drawing and a bridge, for example, is 'the product of mere stipulative fiat' (ibid., p. 75). If this is correct, then there is indeed no special problem about scientific representation. Philosophers of science need no longer occupy themselves with finding even a derivative account of how models represent. It requires only an act of stipulation to bring about an instance of scientific representation. The remaining puzzles may be left to philosophers of mind.

What are we to make of this claim? Callender and Cohen seem to think that it follows directly from the General Gricean position. But it is difficult to see why this should be the case. It is one thing to claim that the representation relation between model and target exists only in virtue of some other, more fundamental form of representation, such as mental representation; it is quite another to claim that an act of stipulation is sufficient to bring about this representational relation. In the first case, we claim merely that *some* form of derivative account of scientific representation may be found; in the second, we commit ourselves to one particular, very simple, form that this account might take. Some of Callender and Cohen's remarks suggest that they take their account not from General Griceanism, but by analogy to the Specific Gricean account of linguistic representation. As we have seen, however, they also explicitly deny that they are committed to Specific Griceanism.

The parallel with depiction is helpful here. Suppose that we were to adopt the General Gricean position with regard to depiction. This would commit us to offering a derivative account of depiction, that is, an account that explained depiction in terms of some other, more fundamental form of representation. In fact, there are rather a lot of existing accounts that we could draw upon here. Two of these accounts were mentioned in Section 1.3.3: Goldman's Platonic account and Walton's make-believe theory. Either of these accounts might constitute the first step in a General Gricean account of depiction; both purport to explain depiction in terms of some other form of representation. Yet the two accounts are very different, and the continuing debate over depiction within the philosophy of art suggests that taking either to constitute the first step in a General Gricean account of depiction would be considered far from trivial.

Moreover, neither Goldman's nor Walton's account parallels the derivative account of scientific representation offered by Callender and Cohen. Presumably, in the case of depiction, such an account would claim that a picture depicts its subject if the painter stipulates that it does and intends to bring about the belief in the viewer that it does. Neither Goldman nor Walton take such an act of stipulation to be sufficient for depiction. And it is clear why they do not, for stipulation is plainly not sufficient for depiction. Suppose we took a blank canvas and stipulated that it represented Napoleon, and that we intended to bring about the belief in others that this canvas represented Napoleon. And suppose further that this intention was recognised and our audience did believe that the canvas represented Napoleon. The blank canvas might, then, be said to represent Napoleon, in some sense. Such is the vagueness of the term 'represent'. And yet clearly it would not *depict* him.

Adopting the General Gricean position with regard to pictures, then, does not commit us to the view that stipulation is sufficient for depiction. Instead, it leaves us with many possible ways of attempting to explain why pictures depict in terms of other, more fundamental, forms of representation. Similarly, we might adopt a General Gricean approach to models without taking stipulation to be sufficient for model-representation. Just as there are many different candidates for a derivative account of depiction, so there might be many different derivative accounts of model-representation. Nevertheless, we might still ask whether the account that Callender and Cohen propose is successful. Is an act of stipulation sufficient for model-representation?

1.3.5 Stipulation and salt shakers

To support their claim that stipulation is sufficient for scientific representation, Callender and Cohen ask us to suppose that we were to pick up a salt shaker and stipulate to our dinner partner that it represents Madagascar. Then, Callender and Cohen point out, as long as our stipulation is understood,

> when your dinner partner asks you what is your favorite geographical land mass, you can make the salt shaker salient with the reasonable intention that your doing so will activate in your audience the belief that Madagascar is your favorite geographical land mass. (ibid., p. 74)

According to Callender and Cohen, this example shows that an act of stipulation, if properly recognised, is sufficient to establish an instance of scientific representation. Is this correct? Would we say that the salt shaker represents Madagascar? In some sense of the term 'represents' no doubt we would; again, the term is loose enough to support many different uses. But would we say that the salt shaker *model-represents* Madagascar? Would it represent Madagascar in the same way that Crick and Watson's model represents the DNA molecule, for example, or Bohr's model represents the atom?[7]

Let us again look to depiction. Perhaps the account of depiction that comes closest to claiming that stipulation is sufficient for depiction is Nelson Goodman's conventionalist account. According to Goodman, the relation between a picture and what it depicts is like that between a name and its referent; both refer to, stand for or denote, their objects. Resemblance or similarity are neither necessary nor sufficient conditions for a picture to denote its object. In fact, 'almost anything may stand for almost anything else' (Goodman, 1976, p. 5). One way to establish denotation, it seems, is by stipulation. If we stipulate that the blank canvas represents Napoleon then the canvas may be said to denote Napoleon. However, even Goodman does not take denotation to be sufficient for depiction. Instead, he recognises that his theory must account for the considerable intuitive differences between pictorial and nonpictorial representations. And he attempts to do so by presenting a number of formal criteria that are intended to distinguish pictorial symbol systems from nonpictorial ones, such as linguistic or diagrammatic symbol systems.

Both David's portrait and the name 'Napoleon' may be said to represent Napoleon. Perhaps both refer to or denote him. Similarly, both Crick and Watson's model and 'D.N.A.' might be said to represent or refer to or denote the DNA molecule. However, any theory of depiction which counted the name 'Napoleon' a *depiction* of Napoleon would have failed to capture something important about the way that David's portrait represents Napoleon. It would have failed to characterise what, intuitively, seems to be so special about pictorial representation. In the same way, it seems that any theory of model-representation that counted 'D.N.A.' a model-representation of the DNA molecule would have failed to capture something important about the way Crick and Watson's model represents the DNA molecule. It would have failed to characterise the particular form of representation that Crick and Watson's model provides.

Of course, our intuitions regarding scientific models are less clear cut than our intuitions regarding pictures. However, there still seem to be many important differences between Crick and Watson's model and the name 'D.N.A.' that our theory must explain. We think that the form of the name 'D.N.A.' is ultimately arbitrary, for example; any combination of letters could have done the job just as well. But the form of Crick and Watson's model was the product of years of research and careful adjustment. Unlike the name 'D.N.A.', Crick and Watson's model seems to 'tell us' something about the DNA molecule, and we feel that in some way what it tells us can be right or wrong, accurate or inaccurate. Scientists base their explanations of biological phenomena on the structure of the DNA model. But few would seek to explain a molecule's behaviour in terms of the letters which appear in its name. Our theory of how models represent must account for these intuitions.

To sum up: Callender and Cohen offer little argument in support of General Griceanism. And yet, even if we were to adopt this approach to explain how models represent we would neither solve nor dissolve the problem of scientific representation. Instead, we would simply be left with the task of providing a derivative account of model-representation, explaining how models represent in terms of some other, more fundamental, form of representation. The particular derivative account assumed by Callender and Cohen, which claims that models represent simply in virtue of an act of stipulation, is clearly inadequate. In fact, their argument simply trades on the ambiguity of the term 'represent'. An act of stipulation may perhaps be sufficient to make it such that a model refers to or denotes some system, and so 'represents' it in some sense of the term. But, just as theories of depiction aim to account for the particular form of representation that pictures provide, so our theory of scientific representation should account for the particular form (or forms) of representation offered by scientific models. Stipulation is not sufficient to establish an instance of this relation.

1.4 Conclusion

In this chapter, we have encountered two main problems concerning scientific modelling. The first focuses on theoretical modelling: what are model-systems and how are we to understand scientists' discourse about them? The second problem exists for both theoretical and physical modelling, and asks how scientific models represent the world. Contrary to Callender and Cohen, we saw that this problem cannot be dismissed by philosophers of science, even by those who believe that scientific representation may be reduced to some more fundamental form of representation.

We will return to the question of how models represent in Chapter 3. First, however, we must see whether the make-believe view can help us solve the ontological problems posed by theoretical modelling.

2
What Models Are

It is time to introduce the make-believe view and see if it can help us to understand what is going on in scientific modelling. In this chapter, we will focus on ontology. Section 2.1 will provide a brief introduction to the central features of Walton's theory. We will then see how this theory may be applied to both physical and theoretical models (Sections 2.2.1 and 2.2.2). In the remainder of the chapter, we will focus on the ontology of theoretical modelling. In Section 2.3, we will see that the make-believe view can make sense of what scientists are doing when they model the world without positing any object that satisfies their modelling assumptions. In Section 2.4, we will examine how this direct account of theoretical modelling differs from those that compare model-systems to fictional characters.

2.1 Walton's theory: representation and make-believe

2.1.1 Props and games

The central notion in Walton's theory of art is that of a *game of make-believe* (1990, pp. 35–43). Suppose some children play a game in the woods in which they imagine tree stumps to be bears. In Walton's terminology, in this game, the tree stumps are *props* and the convention that the children establish by their agreement that stumps 'count as' bears is a *principle of generation*. Together, props and principles of generation make propositions *fictional*. To say that a proposition is fictional, in Walton's sense, is to say that there is a prescription to imagine it. Thus, given the rule that stumps 'count as' bears, if a participant in the game comes across a stump in a thicket, she is to imagine that a bear is there. The presence of the stump, together with the principle of generation, makes it fictional that a bear is in the thicket. The fact that this is fictional is said to be a *fictional truth*.

What is fictional in a game of make-believe need not be the same as what is imagined. A stump which remains hidden under a pile of leaves still makes it fictional that a bear lurks there, even if this is never imagined by anyone playing the game. An oddly shaped stump might prompt one of the participants to imagine a wolf and not a bear, but unless the game is changed, the proposition that there is a wolf before them is only imagined, not fictional. Fictional truths therefore possess a certain kind of 'objectivity'. Participants can be unaware of fictional truths and mistaken about what is fictional. Finding out what is fictional may be very difficult: in order to find out whether it is fictional that a bear is hiding under some leaves, the children must actually dig down and look. Simply imagining that there is a bear hiding there is not enough to make it fictional that there is. What is fictional in a game of make-believe depends only on the props of the game and the principles of generation in effect.

Sometimes, when we call something 'fictional' we do so to imply that it is false or even deceitful. It is important to note that Walton does not use the term in this way. To say a proposition is fictional in Walton's sense is simply to say that there is a prescription to imagine it. This is perfectly compatible with truth. If a child screams when he comes across a stump in the woods, it will probably be fictional that he screams; it is both fictional and true that the child screams. (We will consider how this relates to other notions of fiction in Chapter 3).

2.1.2 Representations

Walton uses his framework to develop a theory of representation: representations, he suggests, 'are things possessing the social function of serving as props in games of make-believe' (1990, p. 69). And he argues that this concept of representation applies to many novels, paintings, sculptures, plays, films and musical works. Many other entities that we might normally call 'representations', however, such as most history books, newspaper articles, biographies or textbooks, Walton thinks, do not count as representations in his sense. The function of a biography of Napoleon, it seems, is not to prescribe imaginings about Napoleon, but to make certain claims about him. The biography does ask us to believe certain things of Napoleon, and it is arguable that believing something requires us to imagine it. But there is no rule that we ought to believe what the biography says about Napoleon simply because it says it. On the other hand, Walton claims, there is a rule that we ought to imagine certain things of Napoleon when we read *War and Peace*, simply because the novel is written as it is. For this reason, the novel counts as a representation in his sense.

The stumps in the children's game are not representations. In general, it is not the function of tree stumps to serve as props in our society, even though they may be used as such. And there are, of course, further important differences between the stumps and representations such as novels, paintings and plays. Two points should be noted in particular. The first concerns principles of generation. Often, the rules of children's games are declared explicitly ('Let's pretend the tree stumps are bears!'). But principles of generation need not be explicitly defined, and the rules guiding our imaginings with most works remain implicit. The second point that should be noted concerns what Walton calls *reflexivity* (ibid., pp. 117–21). As well as prescribing imaginings, the stumps are also the objects of those imaginings: the children imagine that the stumps themselves are bears. This is not a necessary condition for something to count as a prop, however. The text of *War and Peace* may ask us to imagine things about Napoleon, but clearly, we are not to imagine that the page of text itself is Napoleon. Nevertheless, some works of fiction are reflexive. For example, when we read the first chapter of *Dracula*, we are to imagine that the text we are reading is an excerpt from the character Jonathan Harker's journal.

Finally, some terminology: *authorised* games are those in which it is the function of a representation to be used as a prop. Of course, representations might be used in different games entirely. For example, a child might use a copy of *Dracula* as an 'island' in a game of toy soldiers. Walton calls these *unofficial* games (ibid., pp. 405–11). Something is an *object* of a representation on Walton's theory if there are propositions about it which the representation makes fictional (ibid., pp. 106–37). Napoleon is an object of *War and Peace*, as is St Petersburg. Salisbury Cathedral is an object of Constable's *Salisbury*

Cathedral from the Meadows. *Representation-as* is a matter of what propositions about an object a representation makes fictional. *War and Peace* represents Napoleon as invading Russia in 1812. In this respect, the novel corresponds to Napoleon. If a representation corresponds completely with its object then it *matches* it (ibid., pp. 108–10). But a work may represent something it does not match and match something it does not represent. It is fictional in *The War of the Worlds* that London is attacked by Martians in the late nineteenth century. The novel represents London, but does not match it. Conversely, a portrait of John may match his twin brother David, but it represents John and not David.

With this outline of Walton's theory in place, let us now begin to apply it to scientific models.[1]

2.2 Models as make-believe

2.2.1 Physical models

Consider a statue of Napoleon on horseback. How does it represent Napoleon? According to Walton's theory, the statue functions as a prop in games of make-believe. The games authorised for it are governed by certain principles of generation that apply to statues of this type. The principles of generation state what appreciators of the statue are to imagine, given its features. These principles may be difficult to state explicitly. One might be that if the statue has a certain shape, then we are to imagine that its subject has a certain shape. On the other hand, it is not usually a principle for sculptures that if the sculpture is made from marble, then, fictionally, its subject is made from marble. Together, the principles of generation and the features of the statue generate certain fictional truths. Some of these fictional truths are about Napoleon. For example, it will be fictional that Napoleon is riding a horse. It may also be fictional that he is handsome. In Walton's terminology, then, Napoleon is an object of the statue; the statue represents him as handsome and riding a horse. If, in fact, Napoleon was ugly, then we may say that the statue represents Napoleon, but does not match him.

Now consider a metre-long scale model of the Forth Road Bridge. I think that, like the statue, we may regard this model as a representation in Walton's sense. The model functions as a prop in games of make-believe, which are governed by certain principles of generation appropriate for scale models of this sort. Suppose, for example, that the model is built to a scale of 1:1000. Then one principle will be that, if part of the model has a certain length, then, fictionally, the corresponding part of the bridge is a thousand times that length. Together, the features of the model and the principles of generation generate fictional truths. For example, if the model is a metre long, it will be fictional that the bridge is a thousand metres long. The bridge is therefore an *object* of the model; the model *represents* it *as* a thousand metres long. Like the statue, the model represents the bridge but does not *match* it (in fact, the Forth Road Bridge is 1006m long).

This is merely a sketch of how Walton's theory may be applied to physical models. The details will vary widely from case to case. For example, recall the Phillips machine, which represents the economy using tanks of coloured water. Clearly, the principles governing its use will be very different from those that apply to scale models. (One might be that if water is flowing through a certain pipe, then fictionally, taxes are being paid.) Later on, we will see how the make-believe view may be used to give an in-depth analysis of a number of different cases of physical modelling. Now, however, let us turn to theoretical modelling.

2.2.2 Theoretical modelling

Consider the example of theoretical modelling introduced in Chapter 1. When we model the bob bouncing on the end of a spring as a simple harmonic oscillator, we take the bob to be a point mass m subject only to a uniform gravitational field and a linear restoring force exerted by a massless, frictionless spring with spring constant k attached to a rigid surface. This is our prepared description of the bouncing spring. We also take the spring's motion to be governed by the equation of motion: $md^2 x / dt^2 = -kx$ (2.1). How are we to interpret the prepared description and equation of motion that we write down?

Here I think we may look to works of fiction. As we saw in Section 2.1, Walton applies his account to literary fiction, like novels or stories, as well as to works like paintings or statues. A novel like *The War of the Worlds* is also to be understood as a prop in a game of make-believe, just like the stumps in the children's game in the woods. The text of the novel, together with the principles of generation appropriate for novels of this sort, generate fictional truths; they prescribe readers to imagine certain things. Of course, the principles that apply to novels or stories are very different from those that apply to the stumps or to statues. They are conditional upon the text of the novel, not on the location of tree stumps or the form of the sculpted marble. As we saw in Section 2.1.2, one notable difference between many novels and stories and the props in the children's game is that, in general, novels are not reflexive representations; they do not prescribe imaginings about themselves.

Keeping these differences in mind, I believe that our prepared description and equation of motion may be understood in the same way that Walton understands works of fiction. Consider the following passage from *The War of the Worlds*:

> The dome of St Paul's was dark against the sunrise, and injured, I saw for the first time, by a huge gaping cavity on its western side. (Wells 1898/2005, p. 170)

Clearly, this is not a description of St Paul's Cathedral. When Wells wrote this he was not claiming that there really was a hole in the side of St Paul's. Nevertheless, in Walton's view, the passage still represents St Paul's; St Paul's is an *object* of *The War of the Worlds*. As we have seen, in Walton's theory, something is an object of a representation if the representation prescribes us to imagine propositions about it. Usually, Walton thinks, when we read a work of fiction that uses proper names, we take ourselves to be prescribed to imagine things of the normal referents of those names. In this view, the above passage represents (the actual) St Paul's, because it requires readers to imagine certain things of St Paul's, namely that it has a large hole in its dome. In Walton's terminology, the passage makes it fictional that St Paul's has a large hole in its dome.

Some will disagree with Walton's view of proper names in fiction. Without taking a stance in this debate, however, I think we may use Walton's analysis to provide an account of our prepared description and equation of motion. We have seen that these are not straightforward descriptions of the bouncing spring. Nevertheless, I believe, they do *represent* the spring, in Walton's sense: they represent the spring by prescribing imaginings about it. When we put forward our prepared description and equation of motion, those who are familiar with the process of theoretical modelling understand that they are to imagine certain things about the bouncing spring. Specifically, they are required to imagine that the bob is a point mass, that the spring exerts a linear restoring force, and so on.

The suggestion, then, is that we regard theoretical models, like physical models, as representations in Walton's sense. Unlike some physical models, theoretical models are not reflexive representations: we do not imagine that our prepared description is itself a point mass or subject to a linear restoring force. Instead, our description and equation prescribe imaginings about the bouncing spring system. The bouncing spring is an *object* of our model; our model *represents* it *as* a point mass, subject to a linear restoring force and a uniform gravitational field. Using Walton's terminology, we may say that our prepared description and equation of motion make it *fictional* that the bob is a point mass, that it is subject to a linear restoring force, and so on. These propositions are all false of the bouncing spring system; our model represents the system, but it does not *match* it. However, as we have seen, in Walton's theory, a proposition can be fictional and still be true: *War and Peace* makes it fictional that Napoleon invaded Russia in 1812. Similarly, our model makes it fictional that the bob has mass m, but it is also true that it has mass m (if we have measured it correctly).

In the case of our model of the bouncing spring, we model an actual, concrete object: the bob and spring in front of us. As we noted in Chapter 1, however, not all models are of this sort. Sometimes, when we formulate a theoretical model, we do not model any actual system; recall the example of models of the ether. Nevertheless, I believe we may still understand such cases using Walton's theory. The prepared descriptions and theoretical laws that the scientists write down still act as props in games of make-believe. They still act to prescribe imaginings; it's just that there is no actual, concrete object that these imaginings are about. In cases like ether models, scientists' prepared descriptions and equations will parallel passages like the one we considered in Section 1.2.3 from *Dracula:*

> The Count smiled, and as his lips ran back over his gums, the long, sharp, canine teeth showed out strangely. (Stoker, 1897/1994, p. 33)

There is no actual, concrete object that this passage represents: Count Dracula is not a real person. Nevertheless, this passage may still be understood as a prop that prescribes certain imaginings. For example, it prescribes us to imagine that there is someone who has long, sharp teeth. Unfortunately, however, as we saw in Chapter 1, we quickly enter deeper waters. For it seems that such passages also prescribe imaginings *about* someone called Count Dracula and that they license us to say things about that person, like 'Dracula has long teeth'. These problems will also arise for models that represent no actual object, like ether models. The descriptions and equations that scientists wrote down when trying to model the ether also seem to prescribe imaginings about an entity that does not exist. And, like the passage from *Dracula*, they seem to license certain claims about that entity, like 'the ether is always at rest'.

We will consider these problems in Chapter 3. For the moment, we may simply keep in mind that some theoretical models (like our model of the bouncing spring) represent actual, concrete objects, while others (like ether models) do not. In both cases, the make-believe view will understand the descriptions and equations that scientists write down as props which prescribe imaginings. Where the model represents an actual object, some of the imaginings prescribed will be about that object.

2.3 Make-believe and the ontology of modelling

2.3.1 Doing without description-fitting objects

In Chapter 1, we saw that theoretical modelling poses a number of ontological problems. We will now see whether the make-believe view can help us to solve these problems. In this section, we will focus on interpreting scientists' prepared descriptions and equations of motion. In Sections 2.3.2 and 2.3.3, we will turn to consider the way in which we learn and talk about theoretical modelling.

For the moment, we will be concerned only with theoretical models of actual objects, such as our model of the bouncing spring. Models that do not represent actual objects, like ether models, also present us with ontological problems. Indeed, as we saw in the previous section, the problems presented by ether models parallel those generated by passages about fictional characters. It will become clear, however, that the problems presented by ether models are very different from the problems presented by models of actual objects, and it is important that they be kept apart. Much confusion can result from a failure to do so.

So let us return to our model of the bouncing spring. How are we to interpret the prepared description and equation of motion that we write down? I have suggested that they function as props that prescribe imaginings. When we model an actual object, like the bouncing spring, our description and equation prescribe imaginings about that object. For example, when we write down our description and equation we ask others to imagine that the bob is a point mass, that the spring exerts a linear restoring force, and so on.

The key point to notice now is that interpreting our description and equation in this way does not commit us to positing any object of which they are true. To use Thomson-Jones's (2010) terminology again, we need not posit any *description-fitting object*. Our description and equation ask others to imagine things about the bouncing spring which are, as a matter of fact, false. But nothing in this requires us to posit any object that does satisfy that description and equation. For example, the model represents the force exerted by the spring as linear. On the make-believe account, it does so by asking us to imagine that the force the spring exerts is linear. In fact, of course, the force is not linear (it will change dramatically if the spring is stretched too much). The important point, however, is that imagining that the spring exerts a linear force (when in fact it does not) is no more problematic than, for example, asserting that the spring exerts a linear force (when in fact it does not). Both simply involve a false proposition: in the first case, we imagine it; in the second case, we assert it. Nothing in this, it would seem, requires us to posit any abstract or fictional spring that really *does* exert a linear restoring force. If there is any problem here, it is certainly not unique to our account of theoretical modelling, but instead arises in every case in which we assert a false proposition, consider one, and so on.

One way that we could be misled here has to do with a common way that we think about the content of works of fiction. In *The War of the Worlds*, London is invaded by Martians and the dome of St Paul's is damaged. In everyday talk, we often describe the content of a work as what is 'true in the story' or 'true in its world'. Thus, we might say that it is 'true in *The War of the Worlds*' that St Paul's is damaged. Similarly, we might say that it is 'true in the model' that the bob oscillates sinusoidally. Some philosophical analyses take this kind of talk about fiction seriously, and do conceive of 'truth in fiction' (i.e., what is part of the content of the work) as a kind of truth. David

Lewis (1978), for example, analyses truth in fiction as what is true in certain possible worlds that are picked out by the story.

All that is important for our purposes, however, is that Walton does not conceive of 'truth in fiction' in this way. As we have seen, in his view, the content of a work is what it makes fictional. Recall that to say that a proposition is fictional is not to say that it is true in some fictional world; it is merely to say that there is a prescription to imagine it. To say that *The War of the Worlds* makes it fictional that St Paul's is damaged is not to say that there is some fictional realm in which it is true that St Paul's is damaged; it is merely to say that the novel prescribes us to imagine this. Similarly, to say that it is fictional that the bob is subject to a linear restoring force is not to say that there is any object of which this is true. It is merely to say that we are to imagine of the actual bob that it is subject to a linear restoring force.

Thus, on the make-believe view, when scientists model a real system there is no object that satisfies their modelling assumptions. This account is therefore very different from that offered by Giere. It also differs from the indirect fictions view. As we saw in Chapter 1, both of these accounts offer an indirect, two-stage view of scientific modelling (Figure 1.1, repeated here as Figure 2.1). By contrast, I propose a direct view. Our prepared description and equation of motion represent the bouncing spring directly, by prescribing imaginings about it (Figure 2.2). There is no intermediate entity of which what we imagine is true. There are no model-systems.

Why have people thought that we need such entities? We encountered one argument in favour of description-fitting objects in Chapter 1, in the following passage from Giere:

> If we insist on regarding principles as genuine statements, we have to find something that they describe, something to which they refer. (2004, p. 745)

| Prepared description and equation of motion | Specify → | Model-system | Represents → | Target system |

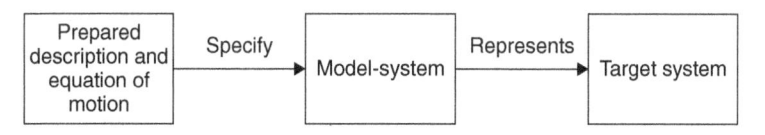

Figure 2.1 Indirect views of theoretical modelling

```
┌─────────────┐                              ┌──────────────┐
│  Prepared   │     Prescribe imaginings     │              │
│description and├─────────────────────────────▶│Target system │
│ equation of │            about             │              │
│   motion    │                              │              │
└─────────────┘                              └──────────────┘
```

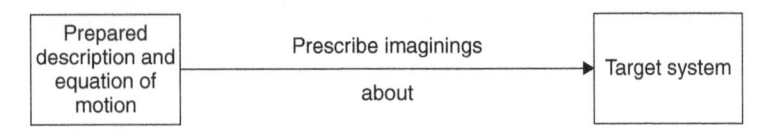

Figure 2.2 Models as make-believe

The correct response to this argument, I think, is simply to deny that we need to regard theoretical principles formulated in modelling as genuine statements. Instead, they are prescriptions to imagine. If theoretical principles are understood in this way then there is no reason to think that there needs to be any object which they describe, any more than we need to posit a new object every time we give a false description of something. And we also need not posit any object to which the principles *refer*; they refer to the actual bouncing spring system in front of us.

The indirect fictions view presents us with a different argument for description-fitting objects. As we saw in Chapter 1, this view compares scientists' prepared descriptions to passages about fictional characters. If we accept this comparison then we might well think we require description-fitting objects. After all, according to realists, we need to posit fictional entities to make sense of passages about fictional characters. This is not the comparison I have drawn, however. Where scientists model an actual object, I have argued, their descriptions and equations parallel passages in fiction that feature actual people or places, like the passage about St Paul's from *The War of the Worlds*, not passages about fictional characters. I will say more about why I think this is the correct parallel to draw in Section 2.4. For the moment, we may simply note that, if we understand our description and equation as I suggest, then problems with fictional characters do not arise. The passage from *The War of the Worlds* does not require us to posit a fictional, damaged, St Paul's; it simply represents the actual St Paul's. Similarly, our description and equation do not require us to posit a fictional, idealised, bouncing spring; they simply represent the actual bouncing spring.

Once again, our common ways of talking could mislead us here. Speaking loosely, we might say that our model of the bouncing spring 'represents' a point mass or a massless spring. Point masses and massless springs are certainly not things that we can collect from the lab store cupboard, and it is tempting to label them as 'fictional entities'. Speaking more carefully, however, we should say that our model represents an (actual) pendulum bob *as* a point mass and an (actual) spring *as* massless and frictionless. For this reason, the model does not give rise to any need to posit fictional entities. The distinction here can be illustrated by looking to pictures. At the 2005 British general election, cartoons were published that depicted Conservative Party leader Michael Howard as a vampire. Like point masses, vampires do not exist. But these cartoons do not present the same problems as the novel *Dracula* or a picture of the Count would. Like our prepared description and equation of motion, the cartoon represents an actual object, namely Michael Howard.

So the make-believe view does not require us to posit any description-fitting objects to make sense of scientists' modelling assumptions. In my account, there are no model-systems. However, as we saw in Chapter 1, scientists sometimes talk as if there were such objects and as if we could learn about their properties. How are we to understand this?

2.3.2 Learning about theoretical models

To see how we can understand learning in theoretical modelling without any description-fitting object, we may draw on Walton's analysis of the *worlds* created by works of fiction. Recall that *authorised* games are those in which it is the function of a representation to serve as a prop. Thus, reading *The War of the Worlds* and imagining Martians invading London is authorised; imagining the words on the page to be works of abstract art is not. What is fictional in the *world* of a representation is that which would be fictional in any authorised game and whose fictionality is generated by the work alone (Walton, 1990, p. 60).[2] Thus, it is fictional in the world of *The War of the Worlds* that London is attacked, that St Paul's is damaged, and so on.

Primary fictional truths are those that are generated *directly* by the props, together with the relevant principles of generation (ibid., pp. 140–4). Thus, the presence of the stump in the clearing, together with the principle that stumps count as bears, generates the fictional truth that a bear is in the clearing. *Implied* fictional truths are generated *indirectly* by other fictional truths (ibid.). *Dracula* makes it fictional that a large dog jumps off a deserted boat that sails into

Whitby harbour. Given what we have read so far, we can infer that this dog is none other than the Count in animal form. Even though the text does not directly say so, then, it is fictional that Dracula jumped off a deserted boat that sails into Whitby harbour. The fact that this is fictional is an implied fictional truth: it is fictional that Dracula jumped off the ship (in part) because it is fictional that a dog jumped off the ship.

Thus, we may divide the principles by which fictional truths are generated into two kinds: *principles of direct generation* and *principles of implication*. The former are conditional upon the features of the representation. They say, for example, that if a novel contains certain words then certain fictional truths are generated. Principles of implication tell us what further fictional truths are implied by primary fictional truths.

This notion of the world of a representation may also be applied to our model of the spring. It is fictional in the world of our model that the bob is a point mass, that the spring exerts a linear restoring force, and so on. These are primary fictional truths: they are generated directly by our prepared description and equation of motion and the principles of direct generation in effect. These principles of direct generation would appear to be rather simpler than those governing the use of novels or paintings. (There appears to be no analogue of the deceitful narrator for scientific models, for example.) When we read a particular prepared description and equation of motion, it seems, we are to imagine that they are true of the system being modelled.

These primary fictional truths in turn generate other, implied, fictional truths. They imply, for example, that it is fictional that the bob oscillates sinusoidally and that it reaches its greatest speed when it passes through the equilibrium position. They also imply that, fictionally, the bob oscillates with a period of $T = 2\pi\sqrt{m/k}$. This is fictional in the world of our model even though we did not write down this equation when we came up with the model. Competent users of the model understand that they are to imagine these things to be true of the bouncing spring, *given* that they are to imagine that it satisfies our prepared description and equation of motion. What principles govern these implications? It is tempting to suggest the principle that if, fictionally, a variable satisfies a certain equation then it should also satisfy its solutions. But often during the course of a calculation, solutions are 'discarded' as having no physical interpretation. (For example, we discard the negative square root for the bob's period of motion.) And, in fact, I believe that principles of implication are more difficult to specify explicitly and will vary from case to case. (To take an obvious example, it seems we make different inferences in models based on classical and quantum mechanics.)

Even without an explicit statement of the various principles of generation, however, this account provides us with a way of understanding learning about a theoretical model. This is not a matter of learning facts about any object. Instead, it is a matter of discovering what is fictional in the world of the model. As we saw in the case of the children's game in the woods, what is fictional in a game of make-believe can be unknown to us: the child may not know that it is fictional that a bear is hiding under a pile of leaves. Similarly, we may not know when we formulate our model that it is fictional that the bob oscillates sinusoidally or that its period of oscillation is $T = 2\pi\sqrt{m/k}$. The reason for our ignorance is not the same as in the child's case, however. Our position is more like that of the reader of *Dracula* who fails to realise that the dog who jumps off the boat in Whitby harbour is the Count. We are quite aware of the state of our props, and of many of the fictional truths these props generate. What we don't know are many other fictional truths that these primary fictional truths imply. Learning about our model is a matter of discovering these implied fictional truths. It is a matter of discovering what further propositions we are required to imagine, once we imagine our prepared description and equation of motion to be true.

The notion of the world of a model also allows us to meet an important objection that might be raised against the make-believe view. One reason for identifying models with model-systems, understood as abstract or fictional entities that satisfy prepared descriptions and theoretical laws, rather than the descriptions and laws themselves, is that it seems that the same model may be given many different formulations. We produce the same model if we write our assumptions in Portuguese or French, for example. The notion of the world of a model allows us to explain this. Just as different linguistic formulations may define the same abstract or fictional object, so they may generate the same fictional world; that is, they may each prescribe us to imagine the same propositions about the bouncing bob.

2.3.3 Talking about theoretical models

As we saw in Chapter 1, much of our talk about theoretical modelling appears to assume there is an object of which our prepared description and equation of motion are true. And yet the account I have proposed denies that there is such an object, even an abstract or fictional entity. If we deny that there are model-systems, how can we make sense of what scientists say in theoretical modelling? In this section, we will consider this question in detail. First, we will consider statements that appear to state what is 'true in our model', like 'the force exerted by the spring is linear' or 'the bob oscillates sinusoidally'. Then we will consider an important example of talk about model-systems that does not fall into this category. These are what Giere calls 'theoretical hypotheses' – that is, statements in which we claim that the model-system is similar to the world in certain respects and degrees.

2.3.3.1 Talk about what is 'true in the model'

When we model the bouncing spring we might say things like 'the bob oscillates sinusoidally', 'the force exerted by the spring is linear', 'the system does not dissipate energy', 'no air resistance acts on the bob' or 'the spring has no mass'. If these utterances are taken to express claims about the actual bouncing spring, they are false and known to be so. And yet if we utter them in the context of our model we would appear to assert something true. If a teacher modelled the spring and asked a student how the bob behaves, and she replied 'the bob oscillates sinusoidally', she would answer correctly. The student appears to make a straightforward assertion, and yet, in my account, there is no object of which this assertion is true. How can we explain this?

I propose the following analysis: when the student says 'the bob oscillates sinusoidally', what she actually asserts is that the prepared description and equation of motion that the teacher writes down make it *fictional* that the bob oscillates sinusoidally. In other words, she claims that, given the description and equation, one is supposed to imagine that the bob oscillates sinusoidally. This statement is true and its truth depends only on the description, equation and the relevant rules of generation; it does not depend upon the existence of any object that really does oscillate sinusoidally.

So understanding talk about what is 'true in our model' does not require us to posit any object of which that talk is true. When we say 'the bob oscillates sinusoidally' or 'the force the spring exerts is linear', we might appear to describe something that does not exist: no real bob oscillates perfectly sinusoidally and no real spring exerts a perfectly linear restoring force. Because of this, it is tempting to compare our statements with utterances about fictional characters, like Count Dracula. Once again, however, I believe that this is the wrong parallel to draw. Our statements about what is 'true in the model' do not parallel talk about Dracula. Instead, it is as if we were to say 'St Paul's is damaged' while reading *The War of the Worlds*. Utterances like 'Dracula sucks blood' are problematic since they employ empty names like 'Dracula'. On Walton's analysis, the utterance 'St Paul's is damaged' does not contain an empty name: 'St Paul's' takes its normal referent. For this reason, such talk presents no problems with fictional entities. When we say 'St Paul's is damaged' we do not talk about any fictional character; we simply claim that the novel tells us to imagine that St Paul's is damaged.

At this point, we might ask, if what we really assert when we say 'the bob oscillates sinusoidally' is that our model makes it *fictional* that the bob oscillates sinusoidally, why do we make the assertion in this way? Why do we indulge in talk about model-systems? Why not simply say 'it is fictional that the bob oscillates sinusoidally'? Walton suggests one answer. Along with his account of representation, Walton offers an analysis of discourse about fiction, which rests on the idea that we *participate verbally* in games of make-believe (1990, Chapter 10).[3] One aim of this analysis is to deal with statements like 'Dracula has long teeth' in a way that does not commit us to fictional entities. But it may also be used to understand statements without any apparent reference to fictional entities, such as 'St Paul's is damaged'.

According to Walton, such utterances are acts of *pretence*. We pretend to assert that St Paul's is damaged.[4] In doing so, we indicate that pretending in this way is appropriate in games authorised for Wells' novel. It is appropriate because, when we pretend in this way, fictionally, we speak the truth. The reason we, fictionally, speak the truth is, of course, simply that it is fictional in *The War of the Worlds* that St Paul's is damaged. On Walton's analysis, then, when we say

'St Paul's is damaged' we indicate that pretending in this way is appropriate, and in doing so we assert that the conditions that make this pretence appropriate are in place. That is, we assert that it is fictional in *The War of the Worlds* that St Paul's is damaged.

Walton's analysis provides us with a way of understanding utterances about what is 'true in our model', like 'the bob oscillates sinusoidally' or 'the force exerted by the spring is linear'. When we make these utterances we are engaging in (or perhaps merely specifying) certain acts of pretence. When the student says 'the bob oscillates sinusoidally', for example, she pretends to assert that the bob oscillates sinusoidally. In doing so, she indicates that pretending in this way is appropriate in games authorised for the teacher's model. It is appropriate because the model makes it fictional that the bob oscillates sinusoidally. The student asserts that this is the case by pretending in the way that she does. She 'goes along with' the model in order to show us how it works.

Talk of pretence may seem out of place in the context of scientific modelling. Partly, I think, this is due to associations of pretence with deception. But pretence need not be deceptive, of course. Even in the context of a sober academic lecture on *The War of the Worlds*, the speaker might say 'St Paul's is damaged' in order to assert that this is fictional in the novel. We might easily recognise this as pretence without there being any question of our being deceived. When the nature of the pretence involved is seen clearly, I believe this account of discourse in theoretical model is a highly intuitive one. In Chapter 1, we saw that Sklar describes theoretical modelling as the process of 'treating the systems in [a] "pretend" way' in which they are described '"as if" they were systems of some more familiar kind' (2000, p. 71). And in Chapter 5, we will see that verbal pretence is merely one part of scientists' much broader imaginative 'participation' in scientific modelling alongside acts of visual and tactile imaginative engagement with models.

2.3.3.2 Comparing models with the world

Each of the statements I have considered may be described as statements about what is 'true in our model'. Some statements we make while modelling a system cannot be understood in this way, however. A good example of this is the 'theoretical hypotheses' that Giere takes to be important for the way that models are used to represent the world. These are statements of the form 'the system is similar to the model in certain respects and degrees' (e.g., Giere, 2004). Even if we reject Giere's view of representation, scientists do make statements like these. And yet this way of talking about theoretical models also seems to rely on there being objects of which our modelling assumptions are true, since when we talk in this way we appear to compare such an object with the world. We often talk about works of fiction in the same way. When reading *The War of the Worlds* we might say 'before it is destroyed, Wells' London is just like the real city'. We do the same with portraits. For example, while standing before Jacques-Louis David's *Napoleon Crossing the Saint Bernard*, we might observe with a smile that 'David's Napoleon is taller than the real Napoleon'.

The analysis I offered in the previous section cannot be carried over unchanged to theoretical hypotheses. Despite this, I think, we may still analyse theoretical hypotheses without commitment to any object that fits our prepared description and equation of motion. Suppose we were to say 'the period of oscillation of the bob in the model is within 10 per cent of the period of the bob in the system'. When we say this, we are simply comparing what our model asks us to imagine with what is true of the system. Specifically, we assert that the period of oscillation of the bob has some value T_0 and that it is fictional in our model that the bob oscillates with period T_1, where T_1 is within 10 per cent of T_0.

In the previous section, I suggested that Walton's pretence account of discourse about fiction might provide an explanation for why we speak about 'truth in a model' in the way that we do. Pretence also can be used to provide an explanation for the way that we express comparisons between our model and the world in theoretical hypotheses. But the explanation we must offer is a little more problematic. One reason for this is that the pretence must now be understood to occur within an *unofficial*, rather than an authorised, game of make-believe.

On Walton's account, much of our talk about works of fiction is to be understood as invoking unofficial games of make-believe (1990, pp. 404–11). For example, we might remark that, in *The War of the Worlds*, 'Wells laid waste to London'. We often talk in this way, describing authors as bringing about certain events or killing off characters. One way to understand such statements is to suppose that when we say this we invoke a common unofficial game that we play with works of fiction. In this game, to write a novel that makes it fictional that certain events occur is, fictionally, to bring about those events. Even though it occurs within an unofficial, rather than authorised game, we may analyse our utterance in the same way that we analysed talk about what is 'true in the novel'. When we say 'Wells laid waste to London' we indicate that pretending in the way that we do is appropriate in our unofficial game. Our pretence is appropriate because, fictionally, we speak the truth. And the reason we, fictionally, speak the truth is, of course, that Wells wrote a novel in which it is fictional that London is destroyed. We assert that Wells did so by indicating how to pretend appropriately in our unofficial game.

The notion of unofficial games provides us with one way to make sense of comparative statements. For example, consider our utterance, 'David's Napoleon is taller than the real Napoleon'. We may understand this remark as suggesting another common sort of unofficial game, in which it is fictional that there exists both Napoleon and someone called 'David's Napoleon' who, fictionally, has all the properties that the painting attributes to Napoleon. Invoking this game provides us with a simpler, and perhaps slightly more interesting, way of comparing the portrait with the world. When we say 'David's Napoleon is taller than the real Napoleon', we indicate how to pretend appropriately within this game. In doing so, we assert that the conditions responsible for the fact that we pretend appropriately are in place. That is, we assert that it is fictional in the painting that Napoleon is a certain height and that, in fact, Napoleon is shorter than that.

We might analyse theoretical hypotheses in the same way. That is, we might understand theoretical hypotheses as invoking an unofficial game in which it is fictional that there exists both the bob and an entity called 'the model bob' which, fictionally, has all the properties attributed to the bob by the model. By invoking this game, we avail ourselves of a convenient way of making assertions about the accuracy of our model, by indicating how to pretend appropriately in this game. When we say 'the period of oscillation of the bob in the model is within 10 per cent of the period of the bob in the system', we indicate how, fictionally, to speak the truth in our unofficial game. In doing so, we assert that the conditions responsible for our fictionally speaking the truth are in place: that is, we assert that the period of oscillation of the bob has some value T_0 and that it is fictional in our model that the bob oscillates with period T_1, where T_1 is within 10 per cent of T_0.

Unofficial games thus allow us to provide one explanation for why we make assertions about the accuracy of theoretical models in the way that we do. This explanation is not without problems, however. One problem concerns the content of the imaginings prompted within these unofficial games. For example, in the game we play with David's portrait, I said that we imagine that there exists someone called 'David's Napoleon', as well as Napoleon himself. This unofficial game thus appears to create a fictional character. If we wish to understand comparisons between the portrait and Napoleon in this way, then, it seems we will require some account of how it is that we can imagine things about fictional characters. A second problem, raised by Frederick Kroon (1994), concerns the proposition we utter when we say, for example, 'David's Napoleon is taller than the real Napoleon'. In Walton's account, it seems, both 'David's Napoleon' and 'Napoleon' refer to the same thing: Napoleon. If this is the case, then it seems that when we say 'David's Napoleon is taller than the real Napoleon', we pretend to assert a contradiction.

In the end, it may be that both of these problems can be solved. For example, perhaps we will be willing to admit that sentences like 'David's Napoleon is taller than the real Napoleon' express contradictions or, alternatively, to deny that 'David's Napoleon' refers to the real Napoleon. Ultimately, our answers to both of these problems are likely to depend on our wider stance regarding fictional characters. Fortunately, we need not attempt to settle this debate here. Ultimately, it may be satisfying to have an explanation of why we express assertions about the accuracy of models (as well as painting and novels) in the way that we do. But for now, we may remain content with our original analysis, and understand theoretical hypotheses as assertions comparing what our model asks us to imagine about an object with what is true. Nothing in this analysis depends upon the outcome of debates over fictional characters.

2.4 The indirect fictions view

As we saw in Chapter 1, I am not the only person to have suggested that we should understand theoretical modelling by looking to works of fiction. Proponents of the indirect fictions view use analogies with fiction to motivate a very different account of how scientists represent the world in modelling. According to this view, representation in modelling occurs indirectly, through a model-system, and this model-system should be understood in the same way as fictional characters like Count Dracula or Madame Bovary. In the remainder of the chapter, I want to consider this view in depth. As we saw in Chapter 1, the biggest problem facing the view concerns the ontology of fictional characters. In Sections 2.4.2 and 2.4.3, I will examine two solutions that have been proposed to this problem. First, however, in Section 2.4.1, I will explain why I think proponents of the indirect fictions view are wrong to compare all prepared descriptions and theoretical laws to passages about fictional characters.

2.4.1 Direct and indirect views of theoretical modelling

Some theoretical models (like our model of the bouncing spring) represent actual, concrete objects while others (like ether models) do not. I have proposed that our treatment of these different cases should mirror Walton's analysis of fiction. According to this analysis, some passages, like our passage from *The War of the Worlds*, represent actual objects (St Paul's). They do so by prescribing us to imagine things about those objects. Others, like the passage from *Dracula*, do not represent any actual, concrete object but are instead about fictional characters. These passages still prescribe imaginings, but there is no actual, concrete object that those imaginings are about. They therefore give rise to all of the usual problems with fictional characters. Similarly, I argued, when scientists model an actual, concrete object, their prepared descriptions and equation of motion represent that object directly, by prescribing imaginings about it. Where scientists do not model an actual object, then their descriptions and equations are like the passage from *Dracula*. They therefore present us with a host of philosophical problems, analogous to those presented by fictional characters.

The indirect fictions view proposes a very different account. Proponents of this view understand *all* prepared descriptions as if they were like passages about fictional characters, whether the scientist is trying to model a real system or not. In each case, the function of the scientists' prepared description is taken to be that of creating a model-system, which is akin to a fictional character. The only difference between cases in which scientists model an actual system and those where they do not concerns what the scientists do with the model-system afterwards. When scientists model an actual system, they establish another representation relation between the model-system and the world.

Why do proponents of this view believe that we should understand scientists' prepared descriptions in this way? Frigg (e.g., 2010a, p. 257) motivates this view by appealing to the fact that both prepared descriptions and works of fiction contain descriptions of missing systems. But this does not mean that prepared descriptions always parallel passages about fictional characters. Our passage from *The War of the Worlds* is also a description of a missing system: there is no actual, concrete object that satisfies the description it gives of St Paul's. But the passage is not about a fictional character.

Godfrey-Smith points out that 'modelers often take themselves to be describing imaginary biological populations, imaginary neural networks, or imaginary economies' (2006, p. 735). In some cases, I think, this is a good description of what the scientist does. As we will see in Chapter 3, sometimes scientists do not model any actual system, but instead say to themselves 'suppose there were a system like this'. In such cases, we might say that the scientists take themselves to be describing an imaginary system, just as, in writing *Dracula*, Bram Stoker was describing an imaginary vampire. But we must distinguish between these cases and those in which scientists model a real system. When scientists model a real system, I believe, the scientist does not take herself to be describing an imaginary model-system; she simply imagines things about the real system.

To see this, consider Frigg's discussion of the Newtonian model of the solar system. This model makes various assumptions about the solar system that are known to be false, such as that the planets are perfect spheres that interact only with each other. According to Frigg,

> When we read the above description, which tells us to regard the earth and the sun as ideal homogeneous spheres gravitationally interacting only with each other, this description serves as a prop and we engage in an authorised game of make-believe. We imagine the entity described in the description ... We understand the terms occurring in the description and we imagine an entity which has

all the properties that the description specifies. The result of this process is the *model-system*, the fictional scenario which is the vehicle of our reasoning: an imagined entity consisting of two spheres, etc. (2010b, p. 133, emphasis in original)

It is only in the 'next step' that we 'connect our model to the target-system' (ibid., p. 134). We do so by specifying that the model-system denotes the solar system, that 'the sphere with mass m_e in the model-system corresponds to the earth and the sphere with mass m_s to the sun' (ibid.), and so on.

And yet surely it fits more closely with our intuitions to regard the prepared description as prescribing us to imagine things *about the sun and earth themselves*. Intuitively, I think, we regard 'the sun' and 'the earth' in that prepared description to take their normal referents, and refer to the actual sun and earth. And we understand ourselves as being asked to imagine things about the actual sun and earth. After all, Frigg himself writes that the description 'tells us to regard the earth and sun as ideal homogeneous spheres' (ibid., p. 133). Why not avoid excessive philosophical reconstruction and take the description at its word, as asking us to imagine things about the (actual) earth and the (actual) sun?

Another way to put this point is to draw a distinction between two different sorts of imaginings. Sometimes, we imagine people, places and objects that do not exist, like Count Dracula or the ether. Sometimes, however, we imagine things about real objects or people in the world, as when I imagine the walls in my flat painted a different colour, or that I play for Derby County. The mistake that proponents of the indirect view make is to assume that all cases of modelling involve cases of the first sort of imagining. Impressed by the fact that scientists sometimes conjure up imagined systems, just as novelists conjure up fictional characters, they then assume that when scientists represent the world they must do so by somehow using these imagined systems to do so. But another option remains open: the scientist may simply imagine things about the world.

So it is wrong to think that all prepared descriptions parallel passages about fictional characters. Let us put this point aside, however. The indirect fictions view faces a bigger problem. That is the problem of saying what fictional characters are. As we saw in Chapter 1, there is little agreement on this question. Realists think that we can only make sense of passages about fictional characters by allowing that they exist in some sense. They therefore posit fictional entities and offer different accounts of what these entities might be. By contrast, according to antirealists, fictional characters like Count Dracula do not exist in any sense, not even as abstract or Meinongian nonexistent entities. Antirealists therefore attempt to explain how we can understand passages about fictional characters, and our discourse about them, without positing any fictional entities. So it seems that, if they want to shed light on model-systems by comparing them to fictional characters, proponents of the indirect fictions view must tell us how to understand fictional characters. One way to do this, of course, is to look to existing theories of fiction. This is the approach taken by Roman Frigg, who draws on Walton's theory. Let us now turn to consider Frigg's account.

2.4.2 Make-believe and the indirect fictions view?

As we have seen, Frigg (2010a, 2010b) proposes a version of the indirect view of theoretical modelling. In his account, prepared descriptions give rise to model-systems, and these model-systems are 'akin to characters and places in literary fiction' (2010b, p. 100). Frigg acknowledges, however, that without a theory of fictional characters 'explaining model-systems in terms of fictional characters amounts to explaining the unclear with the obscure' (2010a, p. 256). It is for this reason that he looks to Walton's theory. Like me, Frigg proposes that we understand scientists' prepared descriptions as props in games of make-believe. In his view, however, these descriptions do not prescribe imaginings about actual objects; instead, they ask us to imagine a model-system. Frigg calls the relationship between the prepared description and the model-system 'p-representation' (ibid., p. 264). When scientists represent an actual object, they must establish a second representation relation between the model-system and the world, which he calls 't-representation' (ibid.). Frigg offers an analysis of t-representation by drawing on analogies with maps (2010b).

Frigg's aim, then, is to flesh out the indirect fiction view by drawing on an existing theory of fictional characters. The choice of Walton's theory for this task is a little surprising, however. The reason it is surprising is that Walton is an antirealist about fictional characters. In his view, works of fiction may seem to ask us to imagine things about people like Count Dracula or Madame Bovary, and we may seem to be able to talk about them. But, according to Walton, there simply are no such things, not even as abstract or Meinongian nonexistent entities. So if we were to understand model-systems in the same way that Walton understands fictional characters then it seems that we would conclude that there are no model-systems. Following Walton, we would say that prepared descriptions seem to ask us to imagine things about model-systems, and we seem to make claims about them. But in fact, there are no such things, not even as abstract or nonexistent entities.

Frigg wants to follow Walton in his antirealism (e.g., 2010a, p. 264; see also 2010b, p. 120). An antirealist stance on model-systems is difficult to reconcile with Frigg's overall account of theoretical modelling, however. We saw above that model-systems have a central place in that account. In Frigg's view, scientists use model-systems to represent real systems (t-representation). According to his account of t-representation, a model-system X represents some real target system Y if and only if X denotes Y and 'X comes with a key K specifying how facts about X are to be translated into claims about Y' (2010b, p. 126). This might involve, for example, specifying 'object-to-object correlations', such as that which we encountered earlier: 'the sphere with mass m_e in the model-system corresponds to the earth and the sphere with mass m_s to the sun' (ibid., p. 134). Once we have specified such correlations,

> we can then start translating facts about the model-system into claims about the world. For instance, calculations reveal that the model-earth moves on an ellipse, and given that the model-system is an ideal limit of the target we can infer that the real earth moves on a trajectory that is almost an ellipse. (Ibid., p. 135)

If taken literally, however, all of these claims about t-representation would seem to be inconsistent with antirealism. If there are no model-systems then there can be no facts about them and we cannot establish an object-to-object relation between model-systems and the world. If there is no model-earth then it cannot move on an ellipse.

One way to reconcile Frigg's account with antirealism would be to offer some antirealist reinterpretation of what Frigg says about t-representation, which explains away the apparent commitment to fictional entities. If we were to take this route, however, all talk of using model-systems to denote real systems, or of specifying object-to-object correlations between the two, would now be construed merely as a way of talking, rather than as offering an account of how modelling actually works. Another option would be to abandon antirealism. Frigg suggests that he is open to this possibility (e.g., 2010b, p. 113).[5] The problem, of course, is that if Frigg were to reject antirealism, and grant that we must posit fictional entities to serve as model-systems, he would need to provide an account of what fictional entities are. And drawing on Walton's theory will not help to provide such an account.

2.4.3 Deferring the problem?

So the key challenge remains: can proponents of the indirect fictions view flesh out the comparison between model-systems and fictional characters by providing a coherent account of what fictional characters are? As we saw in Chapter 1, however, some have argued that this challenge need not be met. In fact, they claim, worries about the ontology of fictional characters need not concern philosophers of science. For example, in his recent work, Ronald Giere grants that model-systems and fictional characters are ontologically 'on a par' (2009, p. 249). But he questions 'whether we, as philosophers of science interested in understanding the workings of modern science, need a deeper understanding of imaginative processes and of the objects produced by these processes' (ibid., p. 250). Peter Godfrey-Smith appears to endorse a similar attitude. Although he acknowledges that we might need an account of the ontology of fictional characters 'for general philosophical reasons', as philosophers of science, he seems to suggest, we might remain content with accepting the 'folk ontology' of scientific modelling (2006, p. 735).

I am sympathetic to this attitude. As we shall see in Chapter 3, I will also suggest that philosophers of science may legitimately defer many questions about fictional characters to those working in aesthetics, or to philosophy of mind and language in general. The important point to notice, however, is that this route is not open to those who defend an indirect view of theoretical modelling, as Giere and Godfrey-Smith do. The indirect fictions view gives fictional characters a central place in theoretical modelling: in this view, scientists only represent the world in modelling *via* fictional characters. To understand scientific representation, we must therefore understand the relationship between a fictional character and the world. It is difficult to see how we could understand how such things represent without first understanding what they are.

For example, both Giere and Godfrey-Smith describe representation in modelling as a matter of similarities or resemblances between model-systems and the world. If their accounts are to be taken literally, then this will clearly place constraints on the account of fictional characters we can adopt: it must be a realist account, in which there are fictional entities and these entities can possess properties such as mass or velocity. If we wanted to adopt a different view of fictional characters then talk of similarity or resemblance between model-systems and the world would have to be reinterpreted radically. And our account of representation would be reinterpreted along with it. If defenders of the indirect view wish their accounts of scientific representation to aspire to truth, rather than being merely convenient stories, then it seems that they cannot leave fictional characters to philosophers of fiction.[6]

2.5 Conclusion

In this chapter, we have begun to see how the make-believe view may help us to understand scientific modelling. In Section 2.2, we saw that this theory may be applied to both physical and theoretical models. But our main focus in this chapter has been on theoretical modelling and, in particular, on its ontology. I suggested that, when scientists produce a theoretical model of a system, they ask us to imagine that the assumptions they make are true of that system. In Section 2.3, we saw that understanding theoretical modelling in this way does not require us to posit any object that satisfies the scientists' modelling assumptions. Finally, in Section 2.4, we have seen how this direct account of theoretical modelling differs from those that compare model-systems to fictional characters.

Let us now turn to the second key problem that we encountered in Chapter 1: the problem of scientific representation.

3
How Models Represent

Chapter 2 focused on the ontological puzzles raised by theoretical modelling. We will now turn to the problem of scientific representation. First, we will see how the make-believe view may be used to offer an account of representation which meets the criteria set out in Chapter 1 (Section 3.1). This account will draw a parallel between scientific models and works of fiction. In Section 3.2, we will pause to consider some objections to this comparison. Finally, in Section 3.3, we will focus on a type of model that presents problems for theories of scientific representation. These are models which represent no actual, concrete object. As we will see, unlike existing accounts, the make-believe view is able to make sense of these models.

3.1 Make-believe and model-representation

3.1.1 The account

In Chapter 2, I proposed that we understand scientific models as representations, in Walton's sense. Models, I argued, function as props in games of make-believe. In physical modelling, the prop is a physical object, such as the architect's scale model of the bridge. In theoretical modelling, like our model of the bouncing spring, the prop is usually a prepared description and set of equations. In some cases, the prop might be a diagram or picture. Just as for novels or paintings, the principles of generation governing the games in which these props function are complex and vary from case to case. In each case, however, the model represents in virtue of prescribing imaginings. We may formulate this account as follows:

> *MM:* M is a model-representation if and only if M functions as a prop in a game of make-believe.

In saying that M 'functions' as a prop, I mean that it is the social function of M to be used in this way, within the relevant community of model-users. As we have seen, something is an object of a representation, in Walton's theory, if the representation prescribes imaginings about it. According to *MM*, then, a model M will represent a target system T if M prescribes imaginings about T within a game of make-believe. Notice, however, that according to *MM*, it is not a necessary condition for model-representation that the model prescribes imaginings about any target system T. We shall see the importance of this feature of the account in Section 3.3.

In Chapter 1, we coined the term 'model-representation' for the form of representation scientific models employ. We noted that there might be many different forms of model-representation and that each of these forms of model-representation might not be unique to scientific models. An account of representation for scientific models should provide us with conditions that are both individually necessary and jointly sufficient to establish an instance of each form of model-representation that we identify. *MM* is intended to provide an analysis of model-representation. Tentatively, I suggest that the analysis it provides applies to all cases of physical and theoretical modelling. As we have seen, Walton argues that his theory applies to novels, paintings, plays and films. If he is correct, then according to *MM*, model-representation turns out not to be unique to scientific models, but an instance of a much wider form of representation also found in works of fiction. Since some will object to this comparison, we will consider the relationship between models and works of fiction in more detail in Section 3.2.

In the terminology I introduced in Chapter 1, *MM* is a *derivative* account of model-representation: it aims to show how the representational power of scientific models derives from the representational power of certain mental states, namely those of the imagination. The account claims that scientific models represent in virtue of the acts of imagination they prescribe. When a model represents an actual object, it does so in virtue of the acts of imagination that it prescribes about that object. For example, the scale model of the Forth Road Bridge represents the bridge in virtue of the fact that it prescribes us to imagine that the bridge is a certain shape, length and so on. Our model of the bouncing spring represents the spring in virtue of prescribing us to imagine that it is a point mass, subject to a linear restoring force and a uniform gravitational field, and so on.

Notice that the same prepared description and equation of motion may serve very different representational functions. We might write down the same equation and description to represent a different spring system from the one in front of us, for example. Similarly, it seems we could use the Phillips machine to represent the British economy, or the Dutch or the French. Here we encounter a difference between models and works of fiction. Perhaps some works of fiction also represent different objects at different times. We might, for example, imagine a storyteller who travels from group to group telling the same story about 'the leader of the village', 'the oldest in the group', and so on. Another case might be the 'identikit' pictures police use to identify criminals. In general, however, it seems that the object (or objects) of a work of fiction are usually fixed: paintings or novels rarely represent different objects at different times.

The account I have outlined can be amended easily to accommodate this feature of models. As we saw, Walton defines authorised games as those in which it is the function of a representation to serve as a prop, where the function of a prop is determined by the rule adopted by a particular society for using props of that kind. We may accommodate this feature of models, then, simply by allowing that one and the same prop may have more than one function. In the case of the Phillips machine, for example, one of its possible functions might be that of representing the British economy, and another that of representing the French economy. Each of these functions determines a different set of authorised games. In one of these games, the movements of water in the pipes of the machine will be understood to prescribe imaginings about the workings of the British economy, and in another about the French.

3.1.2 Make-believe and salt shakers

In Chapter 1, I argued that we may accept Callender and Cohen's arguments in favour of adopting a derivative account of scientific representation, while rejecting their claim that stipulation is sufficient for scientific representation. Just as there are many different derivative accounts of depiction, there might also be many different derivative accounts of model-representation. *MM* offers one such account. And, unlike Callender and Cohen's stipulation view, *MM* is able to distinguish model-representation from cases of mere denotation or reference.

Recall the example of the salt shaker. Callender and Cohen claim that an act of stipulation is sufficient to make a salt shaker a representation of Madagascar. I argued that, while stipulation might be sufficient for denotation, it is not sufficient for model-representation. According to *MM*, in order to be a model-representation of some object, a model must not only denote that object, there also must be an understanding amongst those who use the model that various imaginings are prescribed that depend upon the features of the model. This is absent in the case of Callender and Cohen's salt shaker. The act of stipulation that they describe may establish that the salt shaker refers to Madagascar, but there is no understanding amongst the diners that they are to imagine anything about Madagascar, given the properties of the salt shaker. For the same reason, my account is also able to exclude names: no convention exists such that we are to imagine certain things of the DNA molecule depending upon the properties of the name 'D.N.A.', such as the number of letters it has or what font it is written in.

In Chapter 1, we noted some important differences between models and merely denoting entities, like names. The form of a name like 'D.N.A.' is ultimately arbitrary, while that of a scientific model is often crucial to its representational function. Furthermore, we noted that scientific models seem to 'tell us' something about their objects, while names do not, and that what the model tells us can be right or wrong, accurate or inaccurate. We are now in a position to explain these differences. The reason that the properties of a model are important to its representational function, while those of names or Callender and Cohen's salt shaker are not, is that the imaginings the model prescribes about its object are conditional on those properties. What a model tells us about its object is dependent on the content of those imaginings, and what it tells us is right or wrong depending on whether the propositions it asks us to imagine are true or false of that object. To put it another way, unlike merely denoting entities, models involve *representation-as* (see also Hughes, 1997). And, on this account, what a model represents its object *as* is a matter of what imaginings it prescribes about it.

Under certain circumstances, the salt shaker could become a model-representation of Madagascar. For example, we might imagine the shaker being used to indicate the location of Madagascar with respect to Africa (denoted by the dinner plate). In this case, the salt shaker (together with the dinner plate) would constitute a model-representation according to *MM*: the salt shaker's properties prescribe us to imagine something about Madagascar, according to rules such as 'if the shaker is to the right of the plate, you are to imagine that Madagascar is to the east of Africa'. One way to establish this rule would be to state it explicitly. As we have seen, however, principles of generation need not be stated explicitly. Many suggest themselves to us almost 'automatically'. Once we have explicitly specified that the salt shaker denotes Madagascar and the plate denotes Africa, it is almost inevitable that we will associate the relative positions of the salt shaker and the plate with the relative geographic positions of Madagascar and Africa. Most participants at the dinner party will take this rule to be in place implicitly, in the absence of any further instructions. The ease with which we understand such conventions, however, should not mislead us into neglecting their importance. No familiar principles of generation come to mind when we are told that the salt shaker represents Madagascar. (Its shape does not readily suggest taking it to be a map of Madagascar, for example.) In the absence of such principles, the salt shaker fails to model-represent Madagascar and merely refers to it; its properties are irrelevant to its representational function, and it can tell us nothing about Madagascar.

3.1.3 Make-believe, misrepresentation and realism

In Chapter 1, I argued that our theory of model-representation should be able to accommodate inaccurate (or incorrect or unrealistic) models as well as accurate ones. The account I have offered meets this criterion. According to *MM*, a model represents an object if it generates fictional propositions about that object. Once again, recall that in Walton's sense, propositions can be fictional and still be true. For example, our model of the bouncing spring makes it fictional that the bob has mass m and that it is attached to a spring, and these statements are both true. However, it is not a condition for model-representation in my account that all, or even any, of the propositions that a model makes fictional must be true. For this reason, *MM* is able to accommodate inaccurate (or incorrect or unrealistic) models. Our model still represents the bouncing spring, even though much of what it asks us to imagine about it is false: the bob is not a point mass, the spring is not massless, and so on. (In Walton's terminology, the model represents its object but does not *match* it.) Or again, like their final double-helical model, Crick and Watson's early models represent the DNA molecule because they prescribe us to imagine things about the molecule. It is simply that some, or even all, of what the early models ask us to imagine is false.

However, the accuracy with which a model represents a system is often of considerable importance, of course. There are many questions that we might ask in this regard. Can we say anything general about the realism or accuracy of scientific models? If we can, how realistic are scientific models in general and in what respects? Are we justified in believing that scientific models are realistic representations of their objects? In this book, we are mainly concerned not with these questions, but with the prior question of how scientific models represent their target systems. As we noted in Chapter 1, however, it would be desirable if our theory of model-representation provided us with a framework in which to address these questions about realism. The theory of model-representation I have proposed does provide such a framework, but this framework differs from that commonly thought to be involved in modelling.

In most accounts of scientific modelling, accuracy is judged in terms of some form of similarity or fit between a model and the world. For example, as we have seen, Ronald Giere takes model-systems to be abstract objects defined by scientists' prepared descriptions and equations of motion. The accuracy of a theoretical model is then a matter of the similarity between this abstract object and the system in certain respects and to certain degrees. In contrast to indirect views of theoretical modelling, such as Giere's, I have proposed a direct view. According to this account, there is no abstract object (or fictional entity or any other kind of object) that satisfies scientists' prepared description and equation of motion. Instead, the description and equation represent the system directly, by prescribing imaginings about it. However, this account still provides us with a way of understanding the accuracy or realism of a model: put simply, a model is accurate in a certain respect if and only if what it prescribes us to imagine in that respect is *true* of the object it represents.

For example, consider our model of the spring again. Whether this model is accurate or not depends upon whether what it prescribes us to imagine is true or not (or, if our standards are less demanding, perhaps only approximately true). For instance, the model asks us to imagine that the bob oscillates with a time period of $T = 2\pi\sqrt{m/k}$. The model is accurate in its prediction if the bob does in fact oscillate with period $T = 2\pi\sqrt{m/k}$. This view may be applied to physical, as well as theoretical models; as we have seen, on the make-believe view, even physical models prescribe us to imagine propositions about their objects.[1] Notice, however, that the account does not claim that all of the imaginings prescribed by models are propositional. One of the reasons for interest in representational devices like diagrams and physical models is a sense that philosophers have focused too heavily on propositional knowledge. In Chapter 5, we will consider some of the nonpropositional imaginings that three-dimensional, physical models prescribe and see how this helps us to explain what sets them apart from other forms of representation.[2]

Making inferences about the world through modelling usually is taken to be a two-stage process: first, we find out the properties of the model-system, and then we translate this into claims about the system. I will say more about how models are used to learn about the world in Chapter 5. But it is now possible to offer an outline of how this process is understood in the make-believe view. In Chapter 2, I argued that learning about a model does not involve learning facts about any abstract or fictional entity. Instead, it is a matter of discovering what is fictional in the world of the model. We make an inference from the model to the system when we take what is fictional in the model to be true of the system (or, perhaps, approximately true). For example, after some calculations we discovered that it is fictional in our model that the bob's period of oscillation is $T = 2\pi\sqrt{m/k}$. If we think our model is accurate in this respect, then we will infer that this is not only fictional, but also true of the spring itself.

3.1.4 Make-believe and similarity

In Chapter 1, we noted a number of parallels between the problem of explaining how models represent and the problem of explaining how pictures represent. A natural response to the puzzle regarding depiction is the thought that paintings represent their subjects because they are similar to or resemble them. Resemblance theories of depiction have a long history and enjoy considerable intuitive appeal. After all, surely the most striking feature of pictures (unlike, say, names) is that they *look like* what they represent. In the same vein, it might be tempting to try to explain scientific representation in terms of similarity between models and the world: many models share important similarities with the systems they represent, just as many pictures resemble their subjects. It is clearly important for its representation function that Crick and Watson's model has a similar structure to the DNA molecule.

Unfortunately, however, attempts to explain representation in terms of similarity quickly run into problems. Nelson Goodman (1976) famously delivered a forceful critique of resemblance accounts of depiction, while Mauricio Suárez (2003) has argued that parallel problems face similarity accounts of scientific representation. The most straightforward version of the similarity account can be formulated as follows:

Sim: M model-represents T if and only if M is similar to T

where M refers to a nonlinguistic abstract or fictional object (in the case of theoretical models) or an actual, concrete object (in the case of physical models), and T refers to the target system.[3] As Suárez (2003) shows, *Sim* faces serious difficulties. One problem is that similarity is a symmetrical relation, while representation is not. The DNA molecule is as similar to Crick and Watson's model as the model is to the molecule, but the molecule does not represent the model. Another, more serious, problem for the account is that similarity is clearly not sufficient for representation. A spiral staircase might also resemble the DNA molecule in some respects, but it does not represent the molecule. The difficulty here is particularly acute, since similarity is ubiquitous; it seems that almost anything is similar to anything else in some respect or other (Goodman, 1972).

The make-believe account allows us to acknowledge the role that similarities play in modelling, while avoiding the difficulties faced by the similarity account. Principles of generation often link properties of models to properties of the systems they represent in a rather direct way. If the model has a certain property then we are to imagine that the system does too. If the model is accurate, then the model and system will be similar in this respect. For example, viewers of Watson and Crick's model are to take it that, if the model has a certain structure, the molecule also has that structure. If Crick and Watson are right, then the model and molecule are similar in that respect. Notice, however, that the model would represent the molecule even if it turned out that Crick and Watson were wrong, and DNA is not a double helix. Moreover, not all principles of generation are so straightforward. In the Phillips machine, water flowing from one tank to another may mean that taxes are being paid. Talk of similarity seems less useful here, although no doubt there is a sense in which the flow of water is similar to the movement of money. In some cases, similarity seems to play no role at all. Normal ball-and-stick molecular models use colour codes to indicate the atoms in a molecule: white for hydrogen, red for oxygen, and so on. But clearly, these are not claimed to be similar to the colours of the atoms. In any case, the important point is that similarity alone is not sufficient to bring about the similarity relation. The spiral staircase could be used to represent the DNA molecule. But to do so we would have to interpret it in a certain way, understanding some features (such as its structure) as representational and some features (such as its colour or texture) as nonrepresentational. Just as with our earlier example of the salt shaker and the dinner plate, we understand such principles with ease: if a biologist explaining the structure of DNA to us were to point to the staircase, we would probably understand what she meant without any further explanation. Nevertheless, without such principles in place, at least implicitly, the staircase remains simply a staircase and does not represent anything at all.

3.2 Models and works of fiction

Many of the things to which Walton applies his theory, such as novels, plays and films, are central examples of what we call works of fiction; many of the things to which he does not, such as history books and biographies, are deemed nonfiction. And in fact, Walton argues that his theory provides a way of distinguishing fiction and nonfiction: a work is fiction, he claims, if and only if its function is to serve as a prop in games of make-believe. In suggesting that models function as props, then, I am suggesting that we try to understand scientific models using an analysis originally intended for works of fiction. The relationship between models and fiction has recently become the subject of considerable interest in philosophy of science. We saw in Chapter 2 that some have compared model-systems to fictional characters, and there is now a growing body of work that, in one way or another, seeks to draw parallels between models and fiction. Unsurprisingly, there are also those who have objected to this comparison.

Before we consider this debate, it will be helpful to recall the overall form of my argument in this book. My aim is not to argue for the make-believe view by analogy. That is, I do not wish to argue that models are analogous to fiction, and because the make-believe theory is a good theory of fiction, it is also a good account of models. Of course, it might be nice if we could provide a unified theory of representation for works of fiction and scientific models (and perhaps paintings, statues, plays and more else besides). And I have often drawn on parallels with fiction in order to understand what is going on in modelling. Ultimately, however, the make-believe view of models must stand or fall on its own merits; it should be judged according to whether it provides a coherent and plausible account of scientific models, one which is able to address the problems we encountered in Chapter 1. In the end, it might even turn out that the make-believe theory must be rejected as an analysis of fiction, while still providing a good account of scientific models.

The key claim, then, is that expressed by *MM*. What matters for us is whether models may be understood as props in games of make-believe, and thus qualify as fiction *in Walton's sense*. Now, of course, *if* Walton is right in his account of what it takes to be a work of fiction, then *MM* will also commit us to claiming that models are works of fiction. But I do not want to defend Walton's characterisation of fiction here. As I see it, we may remain neutral on this point. As a result, the make-believe view does not commit us to the claim that models are works of fiction.[4] Moreover, even if Walton *is* right in his characterisation of fiction, we still need not deny that there are many important differences between models, on the one hand, and novels, plays and so on, on the other. Instead, we would claim only that models share a common core function with such works, namely that of prescribing imaginings. This would not preclude there being many other functions that certain scientific models possess and other works of fiction do not, such as generating predictions or offering certain kinds of explanations (nor, of course, vice versa). Similarly, to claim that some scientific drawings employ the same mode of representation as cartoons and surrealist paintings, namely depiction, would not prevent us from recognising the enormous differences between these different representations.

Nevertheless, it may still be helpful to consider the relationship between models and works of fiction, if only to dispel some of the sense of unease that can accompany any comparison between the two.

One way in which models and fiction have been related in recent philosophy of science is through the work of Hans Vaihinger and his *Philosophy of 'As If'* (1911/1924). Vaihinger's work was first published in 1911 but it was revived by Arthur Fine in his 1993 article, 'Fictionalism', which aims to relate Vaihinger to recent work on modelling. More recently, Fine's paper has formed the focus for a collected volume entitled *Fictions in Science* (Suárez, 2009). As Fine makes clear, Vaihinger operates with his own definition of fiction. For Vaihinger, fictions are elements of theories that 'contradict' reality and are known to do so. Despite being false, Vaihinger thought that some fictions may be 'virtuous' since they fulfil a useful function and he saw fictions at work in many areas of human inquiry. For Fine, Vaihinger's ideas have particular resonance when it comes to scientific modelling:

> Preeminently, the industry devoted to modeling natural phenomena, in every area of science, involves fictions in Vaihinger's sense. If you want to see what treating something 'as if' it were something else amounts to, just look at most of what any scientist does in any hour of any working day. (Fine, 1993, p. 16)

Modelling is certainly full of fictions, in Vaihinger's sense. As we saw in Chapter 1, it is characteristic of modelling that scientists apply prepared descriptions and theoretical laws which they know to be false. While it might be uncontroversial that models contain fictions in Vaihinger's sense, however, it is much more controversial to attempt to treat models as works of fiction. All agree that models invoke false assumptions; whether they function like novels, plays or films is another question. As Eric Winsberg (2009) has observed, that models involve fictions in Vaihinger's sense certainly does not in itself justify our classifying them as works of fiction. A rough map of England might be inaccurate in many ways, and known to be so, but this would not lead us to file it alongside Tolkien's maps of Middle Earth.

Can theoretical models really be compared to works of fiction, like *Dracula* or *The War of the Worlds*? After observing that Vaihinger does not capture our everyday notion of fiction, Winsberg offers his own definition: nonfiction is offered as a 'good enough' guide to some part of the world, and fiction is not (2009, p. 181). While a rough map may not be entirely accurate, then, it is still nonfiction since it is offered as 'good enough' guide to the part of the world it represents. By contrast, Winsberg suggests, works of fiction are not offered as even rough or approximate guides to any part of the world. An objection immediately arises, however. After all, surely some works of fiction *are* offered as guides to the world. Winsberg himself considers the fable of the grasshopper and the ant. While the grasshopper sings and dances all summer, the ant collects food, and when winter comes, the grasshopper must beg for charity. This fable would seem to be offered as a guide to the world. It offers lessons on the value of hard work and putting something by for a rainy day. Despite this, Winsberg argues that the fable does not challenge his distinction between fiction and nonfiction. This is because the fable is 'a useful guide to the way the world is in some general sense' and not to its 'prima facie representational target' (ibid.), a singing grasshopper and toiling ant. Nonfictions, by contrast, *'point to a certain part of the world'* (ibid., emphasis in original) and are a guide to *that* part of the world.

Thus, Winsberg thinks that it is wrong to characterise entire models, or the idealisations they involve, as fiction simply because they are inexact. He does allow that certain elements in models may count as fiction, however. For example, nanomechanics models fractures in silicon by describing the point of fracture using quantum mechanics and the region immediately surrounding it using classical molecular dynamics. To bring together the descriptions of the two regions, the boundary between them is treated as if it contained 'silogen' atoms, which have some properties of silicon and some of hydrogen. Silogen atoms are fictions, Winsberg argues, since they are not offered as even a 'good enough' description of the atoms at the boundary; instead, they are used so that the overall model will work. This overall model, however, is nonfiction. It is offered as a 'good enough' guide to the behaviour of fractures in silicon.

Silogens are fictional elements within an overall model that is nonfiction. It seems that there will also be entire models that qualify as fiction under Winsberg's definition, however. We have already noted that there are models that do not represent any actual object, and we shall see many more examples in Section 3.3. In some cases, of course, such models might still meet Winsberg's criteria for nonfiction since they were indeed offered as good enough guides to some part of the world; it simply turned out that that part of the world did not exist. But, as we will see, there are also models that do not attempt to represent any actual object. Rather like fables, it seems, some of these models are not offered as guides to some particular part of the world but instead aim to give insight in a more indirect way. Nevertheless, if we adopt Winsberg's distinction between fiction and nonfiction then it seems we will have to classify many scientific models as nonfiction. While they involve false assumptions, many models certainly do aim to give a 'good enough' representation of some part of the world.

However, there are reasons to doubt whether Winsberg's distinction does in fact capture an important difference between fiction and nonfiction. In particular, it seems that not all fiction that teaches us about the world does so in the way that Winsberg suggests. To take some obvious examples, consider a historical novel like Robert Graves' *I, Claudius* or a biographical novel like Thomas Keneally's *Schindler's Ark*. Arguably, such novels represent actual people, places and events. *I, Claudius*, it seems, represents the Roman Emperor Claudius. It also represents other real people (Augustus, Tiberius, Livia), places (the palaces of first-century Rome) and events (the death of Augustus, the succession of Tiberius). Moreover, given that the book is a historical novel (rather than, say, fantasy) it seems that it is offered as a 'good enough' guide to those people, places and events in certain respects and we are entitled to take it as such. Of course, parts of the novel, such as its portrayal of Claudius's private thoughts, are entirely fabricated and acknowledged to be so. But then Winsberg's definition allows that nonfiction may have fictional parts.

Of course, there will be those who disagree with this view of historical or biographical novels. Some will argue that they should be classified as nonfiction, or perhaps as borderline cases. Others will claim that the name 'Claudius' in Graves' novel does not, in fact, refer to the real Claudius. Fortunately, we need not enter into these debates. Ultimately, as we saw earlier, what is important for our purposes is not where to draw the correct distinction between fiction and nonfiction, or on which side of this distinction models lie. Instead, what matters for us is Walton's notion of fiction and whether this may be applied to models. Were Walton's account to assume that fiction cannot represent actual places or events, for example, then it might be difficult to see how his theory could be applied to scientific models.[5] But it does not. As we have seen, something may count as fiction in Walton's sense and still represent real people, places and events. It does so by prescribing imaginings about them; that is, it makes propositions about them fictional. And any, or even all, of the propositions that a work makes fictional may be true. The account also allows that fiction may be used to make assertions. In writing *War and Peace*, for example, as well as making it fictional that Napoleon invaded Russia, Tolstoy may also have been claiming that he invaded Russia. What makes a work fiction in Walton's sense is not whether what it says is true, or whether the author asserts it to be true, but in whether the work functions as a prop in games of make-believe. Recall the examples from Chapter 2. The function of a biography of Napoleon, it seems, is not to prescribe imaginings about Napoleon, but to make certain claims about him. The biography asks us to believe things, but there is no rule that we should believe what the biography says simply because it says it. On the other hand, Walton claims, there is a rule that we ought to imagine certain things of Napoleon when we read *War and Peace*, simply because the novel is written as it is. That is why the novel is fiction in his sense, even if Tolstoy was also asserting that Napoleon invaded Russia.

So *MM* claims only that models are fictions in Walton's sense; it claims only that they function as props in games of make-believe. And this allows that models may represent real objects, and that what they say about those objects may be true and even asserted to be true. Still, one might worry whether the conditions expressed by *MM* are sufficient for scientific representation. After all, don't scientists aim to find out the truth about the world, rather than simply asking us to imagine things about it? We have already seen that scientists do not claim that their prepared descriptions and theoretical laws are true. But perhaps they claim that they are 'good enough'? Not all parts of a model are claimed to be even a good enough description of the system being modelled, however. Silogen atoms offer a good example here. But we should not only focus on the models that make it through to journals or textbooks. In the early stages of research, a scientist might not know whether the different parts of a model are a good enough description of the system, and in the end they may turn out not to be good enough at all. But these early models may still represent the system. And, as we have noted already, where a model represents no actual object, it may be that no part of the model is claimed to be a good enough description of any real system.

To sum up: The make-believe account does not commit us to claiming that models are works of fiction. Instead, it claims only that models are fiction in Walton's sense. What makes a model fiction in this sense is that scientists using the model understand that they are to imagine everything that the model says, simply because it says it. Some of the things a model asks us to imagine may be asserted to be true. Some may be asserted to be merely approximately true, or good enough for the task in hand. Other parts of the model may not be claimed to be even approximately true. The mixture of these elements will be different from case to case. In each case, however, those using the model know that they are to imagine what the model says, whether it is asserted or not. This is the function of scientific models: they prescribe a web of imaginings which the scientist can then go on to explore. When scientists model a real system, their model prescribes imaginings about that system. This is what makes the model a representation of the system (though not necessarily a good one).

Let us now turn to consider another problem that faces accounts of scientific representation. This is the problem posed by models which represent no actual object.

3.3 Models without objects

3.3.1 Problem cases

Many scientific models represent actual, concrete objects in the world. Crick and Watson's model represents the DNA molecule.

Bohr's model represents the hydrogen atom. Our model represents the spring bouncing up and down in front of us. As we have noted already, however, not all models are like this. Nineteenth-century physicists constructed mechanical models of the ether. And yet we now know that the ether does not exist. Ether models therefore represent no actual, concrete object. Let us call such cases *models without objects*.

Models without objects are often overlooked in discussions of scientific representation. When they are discussed, it is typically only models of discredited entities like the ether or phlogiston that are offered as examples. In fact, however, there are a much wider range of cases. Many of these are rather mundane. For example, suppose that the government wants to build a bridge and so invites architects to submit models of their proposed designs. Many of these models will not represent any actual object, simply because they are not chosen and their design is never realised. Or, to take another example, consider scientific experiments. Many experiments create events which may never otherwise occur. A scientist might formulate a theoretical model of such an event before funding runs out and the experiment never takes place. Similarly, chemical models may be used to represent molecules which do not normally occur in nature, but which the chemist plans to synthesise in the laboratory (see, for example, Ramberg, 2001).

In addition to cases like these, there are clearly many models that represent no *particular* object or event. I said above that Bohr's model represents the hydrogen atom. But presumably, it does not represent any particular hydrogen atom (although it might be used to do so). Many scientific models are like this. In some cases, we might think that the model represents a type of object or event. Thus, R.I.G. Hughes elects to 'assume without argument that our concept of denotation allows us to denote a type' and offers Bohr's model as an example (1997, pp. S330–31). However, even allowing that we may make sense of the notion of a model representing a type, there are many models, or uses of models, that cannot be thought of in this way. Once again, the comparison with pictures is helpful. Many pictures would seem to represent types. Examples include encyclopaedia illustrations representing certain species of plants, or the famous diagrams of man and woman on the plaque of the Pioneer spacecraft. But clearly not every picture that fails to represent a particular actual object may be thought of in this way. For example, Vermeer's *The Milkmaid* shows a woman pouring milk from a jug by a window. There is no actual woman that the painting represents, nor does it represent a type of woman. Instead, the painting simply represents a particular fictional or 'imaginary' woman. There are numerous pictures of this sort. As Goodman puts it, 'the world of pictures teems with anonymous fictional persons, places, and things' (1976, p. 26).

Analogous cases exist in scientific modelling. Consider the Phillips machine. The machine could be used to represent some particular, actual economy, such as that of Britain. Alternatively, perhaps it could be used to represent a type of economy. But we could also use the machine simply to represent a fictional or imaginary economy. We might begin by saying 'suppose there were an economy like this...' Similarly, suppose that the prepared description and equation of motion that we write down when we model the bouncing bob were to appear instead in a textbook, written to teach students Newtonian mechanics. In this case, it seems there will be no actual system that the model represents, nor type of system. Instead, the textbook simply says, 'Suppose there were a system that was like this. How would it behave?' Or again, a predator-prey model might simply invite us to imagine a population consisting of two species which reproduce at certain rates, without purporting to represent any real population of foxes and rabbits in the world.

3.3.2 What is the problem?

Models without objects present a challenge for theories of scientific representation because, even though they represent no actual, concrete object, they are still representations.

To see this, it helps to consider pictures again. We take for granted that pictures without actual objects are representations. *The Milkmaid* is undoubtedly still a picture, even if there is or was no actual woman that it depicts. When we stand before the painting, we do not see merely a set of brushstrokes on a piece of canvas, but a woman pouring milk from a jug by a window. Indeed, our experience of the picture depends very little upon whether or not the woman actually exists. We can still stand before the painting and admire her care and concentration in her task, just as we might look at David's *Napoleon Crossing the Saint Bernard* and admire Napoleon's bravery and determination. Any account of depiction must therefore find a way to accommodate such pictures, explaining how they can be depictions even though they depict no actual object.

I think the same is true of models. Consider the architects' models mentioned above, each showing proposals for a bridge design. Suppose that all of these models, both successful and unsuccessful, were put on display after the bridge was built. If we were to inspect the models without knowing which one was chosen, our experience of the unsuccessful models would be very similar to our experience of the successful one. Looking at the models, which might be built from balsa wood, paper or construction kits, and might be a metre or ten metres high, we could still recognise each as representing a bridge to be built across the river, and discuss whether that bridge is ugly or beautiful, flimsy or strong. Our account of scientific representation must therefore find a way to accommodate such models, explaining how they can be representations even though they represent no actual, concrete object.

Some are tempted to try to dismiss the challenge presented by models without objects. For example, Callender and Cohen suggest that we might 'bite the bullet and hold that, in cases where x doesn't exist, agents don't succeed in representing x but merely believe they are representing x' (2006, p. 81, note 11). We must be careful here. Of course, everyone would agree that if the ether doesn't exist then there is a sense in which scientists don't succeed in representing it. This is simply to say that ether models fail to represent any actual, concrete object. The real question is whether ether models are still representations, despite failing to represent any actual object. I have argued that they are, and so they must be included by our account of representation. But can we 'bite the bullet' and deny this?

I think this would be a mistake. First, notice that in many of the cases we have considered, agents do not even *believe* that they are representing an actual object. Not all models without objects are like ether models. If we use the Phillips machine to represent an imaginary economy, for example, we do not believe we are representing any real economy; we know perfectly well that we are not. Second, and more important, denying that models without objects are representational would overlook the obvious differences between such models and nonrepresentational entities. Consider the Phillips machine. Water flowing through the pipes of the machine is still taken to *mean something* by those using the machine (that taxes are being paid, for example), even if the machine is not being used to represent any real economy. By contrast, water flowing through the pipes in my flat is not taken to mean anything. Denying that the Phillips machine is representational amounts to simply ignoring the obvious differences between these two cases.

3.3.3 Existing accounts and models without objects

Most existing accounts conceive of scientific representation as a relation. This includes those that try to explain representation in terms of similarity or isomorphism between models and the world (for discussions of these accounts, see Frigg, 2006; Suárez, 1999, 2003). It also includes Ronald Giere's more sophisticated similarity account, on which scientists use models to represent systems by forming 'theoretical hypotheses' specifying their similarities (e.g., 2004, 2010). R.I.G. Hughes' 'D.D.I. account' is not intended to provide necessary and sufficient conditions for representation. However, Hughes does endorse the maxim 'no representation without denotation' (1997, p. S331).[6] Finally, as we have seen, on Callender and Cohen's view, representation in modelling is a relation established by an act of stipulation connecting a model and its object.

These accounts have trouble accommodating models without objects. For it seems that, in order for two things to be related, they must both exist. As they stand, then, none of the accounts just mentioned can explain why models without objects are representational. An ether model cannot represent in virtue of its similarity or isomorphism to the ether if the ether does not exist, nor could a scientist assert the model and ether's similarities in a theoretical hypothesis. The model also cannot denote or stand for the ether, and we cannot establish a representation relation between the model and the ether by stipulation.

One way to preserve the idea that scientific representation is a relation would be to grant the existence of fictional entities. Ether models might then be taken to be representational in virtue of their relation to a fictional ether. Of course, one problem with this suggestion is that the existence of fictional entities is hardly uncontroversial. But even if we were willing to grant their existence, questions would remain for each of the accounts I have mentioned. As we saw in Chapter 2, realists offer different accounts of what sorts of objects fictional entities are. Without further argument, then, it is far from clear that fictional entities would have the right properties to enter into relations of similarity or isomorphism with models. It is also debatable whether models could be said to denote fictional entities, just as they denote actual, concrete objects. Fictional entities are dependent on representations for their existence in a way that normal objects are not. For this reason, the relation between representations and fictional entities (if there are any) seems very different from that between representations and normal, concrete objects (cf. Walton, 1990, p. 127). Finally, could we say that the ether model is representational because it was stipulated that it represents a fictional ether? If any stipulation occurred, it was surely that the model represents the real ether.

One account of scientific representation that does attempt to accommodate models without objects is Mauricio Suárez's 'inferential conception' (2004). In this view, a representational source A represents some target B 'only if (i) the representational force of A points towards B and (ii) A allows competent and informed agents to draw specific inferences regarding B' (ibid., p. 773). At first sight, then, it seems that the inferential conception also regards representation as a relation. However, Suárez argues that this account can accommodate what he calls 'fictional representation, that is, representations of nonexisting entities', and in fact, he claims that in his account 'there is absolutely no difference in kind between fictional and real-object representation – other than the existence or otherwise of the target' (ibid., p. 770).

How is this supposed to work? Consider an ether model. Even though the ether does not exist, perhaps there is a sense in which the model possesses a representational force 'towards the ether'. The ether model is rather like a description such as 'the only inhabitant of London': both purportedly pick out an object. They simply fail to do so because that object does not exist (cf. Walton, 1990, p. 124). As we have already seen, however, not all models without objects are like ether models. When we use the Phillips machine to represent an imaginary economy we do not attempt, but fail, to represent any real economy. As a result, it is difficult to make sense of the idea that this model possesses a representational force (even a thwarted one) towards any actual, concrete object. Once again, we might try to avoid the problem by granting the existence of fictional entities, and allowing that representational force may point towards them too. But then it would be far from clear that there is no difference between representation of real and fictional entities, as Suárez claims. For, as we have already noted, the relation between representations and fictional entities seems very different from the relation between representations and normal, concrete objects.

3.3.4 Solving the problem

The make-believe view does not take representation in modelling to be a relation. Our account of representation (*MM*) made no reference to any object that is represented. According to *MM*, something counts as a model-representation if it serves as a prop which prescribes imaginings; there need not be any object that it prescribes imaginings about. Indeed, the model need not even attempt to represent any object.

The make-believe view is therefore able to accommodate models without objects. On this view, ether models are representational because they prescribe imaginings within a game of make-believe. For example, they might ask us to imagine that the speed of light is constant or that electromagnetic waves are transverse. The models can do this even if there is no ether. Or consider the Phillips machine, used to represent an imaginary economy. According to the make-believe account, the various pipes and tanks of the machine are representational because they are taken to prescribe imaginings according to certain rules. For example, if water is flowing through a certain pipe, we are to imagine that taxes are being paid. No such rules apply to the pipes in my flat. For this reason, the Phillips machine is representational, even when it is not being used to represent any actual economy.

Things are not entirely straightforward, of course. For, in addition to prescribing many unproblematic imaginings, such as that the speed of light is constant, ether models also appear to prescribe imaginings *about the ether*. For example, they might ask us to imagine that the ether is at rest. Once again, then, we encounter problems with fictional entities, this time for the imagination: how are we to understand the contents of imaginings that appear to be directed towards éntities that do not exist, like the ether? This is certainly a problem. Indeed, it is a general problem for theories of intentionality: how are we to understand mental states (beliefs, hopes, desires and so on) that are about things that do not exist (like the ether,

Count Dracula or Pegasus)? Fortunately, this is not a problem that we need to address. All that matters for our purposes is that we are able to imagine things about objects that do not exist. Nobody doubts that we have this ability. The debate concerns how we are to explain it. For this reason, I think, we may defer the problem to philosophers of mind.

Is it appropriate to defer the problem in this way? In Section 2.4.3, we saw that some proponents of the indirect fictions view also advocate a deferral strategy.[7] Thus, Ronald Giere and Peter Godfrey-Smith claim that model-systems are akin to fictional characters, but suggest that philosophers of science should defer problems concerning the ontology of fictional characters to theories of fiction. I argued that Giere and Godfrey-Smith cannot defer the problems in this way. Why, then, am I entitled to do so?

The reason that Giere and Godfrey-Smith cannot defer problems concerning the ontology of fictional characters is that they endorse an indirect view of modelling. In their view, scientists represent the world *via* fictional characters. As a result, our account of scientific representation becomes dependent upon which view of the ontology of fictional characters we adopt. This is not the case in my account. As we saw in Chapter 2, in the make-believe view, when we represent the world we do so directly, not via any fictional character. When our model does *not* represent any real system it may prescribe problematic imaginings that seem to be about entities that do not exist, and some will argue that we need fictional entities to understand such imaginings. But nothing in our account of scientific representation hinges on the outcome of this debate.[8]

3.4 Conclusion

In this chapter, we focused on the problem of scientific representation. We have seen that the make-believe view offers an account of representation that meets the criteria set out in Chapter 1. This account avoids the difficulties faced by the stipulation and similarity view. It also accommodates cases of misrepresentation and provides us with a framework for understanding realism in modelling. In addition, we have seen that the make-believe view has an advantage over existing accounts of scientific representation, since it is able to accommodate models without objects.

So far our focus has been on giving answers to the general philosophical problems posed by scientific modelling introduced in Chapter 1. We will now see whether the make-believe view may be used to make sense of more detailed case studies of modelling in science.

4
Carbon in Cardboard

In the previous chapters, we have seen how the make-believe view may be developed to provide a general account of scientific models, and how this account allows us to solve some important philosophical problems posed by models. We will now ask whether the make-believe view is supported when we look to the practice of modelling in science.

This chapter examines a case from the history of science, focusing on the work of the Dutch chemist J. H. van't Hoff (1852–1911). Winner of the first Nobel Prize in chemistry, van't Hoff was one of the founders of stereochemistry, the part of chemistry that concerns the spatial arrangement of atoms within molecules. The historian Alan Rocke has recently described the rise of atomism in nineteenth-century chemistry as 'one of the great watershed events in the history of science – *the first truly successful entry by human minds and tools into the realm of the invisibly small*' (2010, p. 335, emphasis in original). The birth of stereochemistry marks an important stage in the development of atomism. As we will see, in the decades prior to van't Hoff's work, many chemists were wary of committing to even the existence of atoms (Sections 4.1.1 and 4.1.2). And yet van't Hoff's 'chemistry in space' not only assumed that atoms exist, it also claimed to show the way that they are arranged within molecules. Van't Hoff claimed that the bonds around a carbon atom are tetrahedral, and used this to build up structures for a range of different organic compounds (Section 4.1.3).

Drawing on recent historical studies (especially Meinel, 2004; Ramberg, 2003; Rocke, 2010), this chapter will consider the role of models in the development of stereochemistry. Despite its bold claims, van't Hoff's work met with surprisingly little opposition from chemists (Section 4.1.4). As we shall see, we may better understand the reception of van't Hoff's ideas by focusing on the cardboard models that he mobilised in their support (Section 4.2). Van't Hoff sent these models to leading chemists to publicise his views, and his later works even included templates that readers could cut out to create their own set. Although their theoretical implications were highly contentious, van't Hoff's models displayed a remarkable similarity with those already employed within an established culture of three-dimensional modelling in chemistry. I will argue that the make-believe view offers a framework with which to make sense of these early chemical models and the important role they played in the development of stereochemistry (Section 4.3).

4.1 Chemistry in space

4.1.1 Isomerism and 'chemical structure'

In 1811, Joseph-Louis Gay-Lussac observed that despite their very different properties, sugar, starch and gum arabic all contain the same proportions of chemical elements. Over subsequent years, chemists discovered further examples of the same phenomenon, and in 1830 the Swedish chemist Jacob Berzelius coined the term 'isomer' for substances which contain identical proportions of elements but exhibit different properties (see Rocke, 1984, pp. 167–74). Attempts to explain cases of isomerism drove much chemical theory throughout the nineteenth century. In the 1860s, the debates reached a broad consensus in the form of August Kekulé's 'structural theory' (Ramberg, 2003, p. 15; see also Rocke, 1993). According to this theory, the different properties of chemical isomers were to be explained by their different 'chemical structures'. The term was coined by the Russian chemist Aleksandr Butlerov:

> To be sure, we do not know what connexion exists between the relative chemical effect of the atoms inside a compound molecule and their relative mechanical positions; we do not even know whether, in such a molecule, two atoms which directly affect each other chemically are in fact situated next to one another, but we cannot deny, putting the concept of *physical atoms* entirely to the side, that the chemical properties of a body are determined in particular by the chemical bonding of the elements which form it. Proceeding

from the assumption that there inheres in each *chemical* atom only a specific limited quantity of chemical force (affinity), with which it participates in the formation of bodies, I would designate this chemical cohesion, or the manner of mutual bonding of the atoms in a compound body, by the name *chemical structure*. (Butlerov, 1861, pp. 552–3, as translated in Rocke, 1981, p. 35, emphasis in original)

The chemical structure of a substance thus indicated the manner in which its elements were combined. As Butlerov made clear, however, a substance's chemical structure was not to be confused with the actual arrangement of atoms inside the molecule. Indeed, even the existence of atoms remained highly controversial. Most chemists were committed to the existence of 'chemical atoms' (Rocke, 1984) or 'portions' (Klein, 2003). These were a *'chemically indivisible unit, that enters into combination with similar units of other elements in small integral multiples'* (Rocke, 1984, p. 12, emphasis in original). However, these chemical atoms were not to be confused with 'physical atoms', that is, atoms in the usual sense of submicroscopic particles. Kekulé wrote that

> the question of whether atoms exist or not has but little significance in a chemical point of view: its discussion belongs rather to metaphysics. In chemistry we have only to decide whether the assumption of atoms is an hypothesis adapted to the explanation of chemical phenomena. (Kekulé, 1867, p. 304)

Though he did 'not believe in the actual existence of atoms, taking the word in its literal signification of indivisible particles of matter', Kekulé nevertheless declared his belief in *'chemical atoms'*, that is, 'those particles of matter which undergo no further division in chemical metamorphoses' (ibid., emphasis in original).

In the period leading up to van't Hoff's work, there was, therefore, a 'sharp epistemic distinction' (Ramberg, 2003, p. 16) between knowledge of chemical structures and knowledge of submicroscopic atoms and their arrangements. While the former might be established through chemical experiments, the latter were not a matter for the chemist:

> it is plain that the manner in which the atoms emerge from a decomposing and changing substance cannot possibly prove how

they are located in the stable and unchanging substance. Although it must certainly be considered a task of scientific research to elucidate the constitution of matter, or, if you will, the location of the atoms, yet we must admit that it is not the study of chemical reactions but rather the comparative study of the physical properties of stable substances that can supply means to that end. (Kekulé, 1858/1963, p. 123)

Chemists employed a number of different devices to represent chemical structure. The most influential was that introduced by Alexander Crum Brown in 1864 (see Figure 4.1). Use of Crum Brown's formulas became widespread by the late 1860s, and they remain in use today (Ramberg, 2003, p. 28). In these formulas, elements are represented by circles drawn around their chemical symbol, and valencies by small dashes emerging from the circle. The saturation of valencies – that is, a bond between two elements – is shown by joining the dashes. When his paper was reprinted the following year, Crum Brown replaced the two dashes with a single line, and today, the formulas are often drawn without circles surrounding the chemical symbols. In line with the remarks of Butlerov and Kekulé, Crum Brown was keen to point out that his graphical formulas should be

> used to express constitutional formulae, and by which, it is scarcely necessary to remark, I do not mean to indicate the physical, but merely the chemical position of the atoms. (Crum Brown, 1864, p. 708)

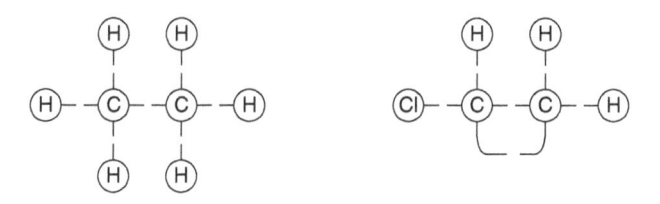

Figure 4.1 Crum Brown's graphical formulas
Source: Crum Brown, 1864, pp. 708–9.

4.1.2 Optical activity and 'absolute isomers'

In 1815, Jean-Baptiste Biot observed that the solutions of some organic substances rotate the plane of polarised light. Such substances are described as 'optically active'. In 1848, in his famous experiments on tartaric acid, Louis Pasteur showed that its optical activity was caused by the presence of asymmetric crystals in the salt; that is, crystals which could not be superimposed upon their mirror images. Both Biot and Pasteur had reasoned that optical activity must somehow derive from an asymmetry in the molecules of active substances. However, by the 1850s and 1860s, the French crystallographic tradition was in decline. Chemists typically used a rather small range of properties to distinguish between substances, such as boiling and melting points, and few measured optical activity. Some even regarded a difference in optical activity alone insufficient reason for classifying two substances as distinct isomers. In 1841, French chemist Charles Gerhardt wrote that:

> Certainly no chemist would dispute the chemical identity of natural camphor and the camphor regenerated by the action of caustic potash on Delalande's oil of camphor, despite the fact that these two camphors differ in rotatory power [that is, their optical activity]. We chemists require chemical differences to distinguish between two bodies. (Gerhardt in *Revue Scientifique*, as cited in Fisher, 1975, p. 33)

One chemist who did make measurements of optical activity was Johannes Wislicenus. In 1869, Wislicenus synthesised a new isomer of lactic acid. Before Wislicenus' work, only two isomers of lactic acid were known: 'milk' lactic acid, derived from fermented milk, and 'meat' lactic acid, derived from muscle tissue. The two had different properties, and were therefore assumed to have different chemical structures. One difference between milk and meat lactic acid, Wislicenus noted, was their optical activity: meat lactic acid rotated the plane of polarised light, while milk lactic acid did not. After Wislicenus' discovery of a third isomer, however, it was clear that milk and meat lactic acid had to have the same chemical structure. The two therefore became an example of 'absolute isomers': isomers whose different properties could not be explained by a difference in their chemical structure. Absolute isomers presented problems for the structure theory. Wislicenus himself thought that the difference would have to be explained in terms of the 'different spatial arrangements of the atoms' (Wislicenus, 1873, as translated in Rocke, 2010, p. 240), but his speculations about the exact form of these arrangements remained rather vague. For example, he suggested that perhaps the atoms of meat lactic acid might not be 'arranged together in the smallest possible space' (Meister, 'Report of Wislicenus', as cited in Ramberg, 2003, p. 46).

4.1.3 Van't Hoff and the tetrahedral carbon atom

Van't Hoff first put forward his theory of the tetrahedral carbon atom in a brief pamphlet of only eleven pages, with the lengthy title *A Proposal for Extending the Currently Employed Structural Formulae in Chemistry into Space, Together with a Related Remark on the Relationship between Optical Activating Power and Chemical Constitution* (1874/2001). The pamphlet was published in Dutch in 1874, and financed by van't Hoff's father. The pamphlet's ambitions are made clear at the outset:

> It is more and more apparent that the current constitutional formulae are incapable of explaining certain cases of isomerism; perhaps this is due to the lack of a more definite pronouncement about the actual arrangement of the atoms. (1874/2001, p. 67)

Van't Hoff proceeded by considering the number of different isomers that would be expected for the various derivatives of methane, on the assumption that the 'four affinities of each carbon atom [are] in four perpendicular coplanar directions' (ibid., p. 67). If this were the way that the atoms were arranged, van't Hoff wrote, then we should expect the number of isomers to be as follows:

> one for CH_3R_1 and for $CH(R_1)_3$;
>
> two for $CH_2(R_1)_2$ (Figures II and III), for $CH_2(R_1R_2)$ and for $CH(R_1)_2R_2$;
>
> three for $CH(R_1R_2R_3)$ and for $C(R_1R_2R_3R_4)$ (Figures IV, V and VI). (Ibid.)

The diagrams van't Hoff referred to in support of his calculations are reproduced in Figure 4.2. These diagrams bear a striking resemblance to Crum Brown's graphical formulas. And yet it is clear that van't Hoff understands them in a very different way than Crum Brown. If the diagrams are taken to show chemical structure, then the positions of the groups around the carbon atom are irrelevant. Van't Hoff, however, took his diagrams to show the actual spatial arrangement of atoms in the molecule.

Van't Hoff rejects a square planar arrangement of atoms, since the number of isomers it would lead us to expect would be 'evidently a much greater number than those known so far' (ibid.). There is, however, a 'second assumption [which] brings theory and fact into agreement' (ibid.). This is the tetrahedral carbon atom: 'by imagining the affinities of the carbon atom directed towards the corners of a tetrahedron whose central point is the atom itself', the number of predicted isomers is only one in the case of molecules with the formulas of the form CH_3R_1, $CH_2(R_1)_2$, $CH_2(R_1R_2)$, $CH(R_1)_3$ or $CH(R_1)_2R_2$, and two for $CH(R_1R_2R_3)$ and $C(R_1R_2R_3R_4)$ (see diagrams VII and VIII in Figure 4.2). That is,

> *in cases where the four affinities of the carbon atom are saturated with four mutually different univalent groups, two and not more than two different tetrahedra can be formed, which are each other's mirror images, but which cannot ever be imagined as covering each other, that is, we are faced with two isomeric structural formulae in space.* (Ibid., p. 68, emphasis in original)

Van't Hoff compares the prediction expected by a tetrahedral arrangement with currently known cases of isomerism. Crucially, however, in addition to those properties regularly employed by chemists to distinguish between isomers, van't Hoff included optical activity. His pamphlet thus provided a list of optically active compounds and their formulas, including lactic and tartaric acid, in order to support his claim that *'every carbon compound that rotates the plane of polarized light in solution contains an asymmetric carbon atom'* (ibid., emphasis in original).

In the conclusion to his pamphlet, van't Hoff drew an analogy between his hypothesis and Pasteur's experiments with tartaric acid: just as Pasteur had related optical activity in the solid state to asymmetric *crystals*, so van't Hoff's hypothesis related optical activity of

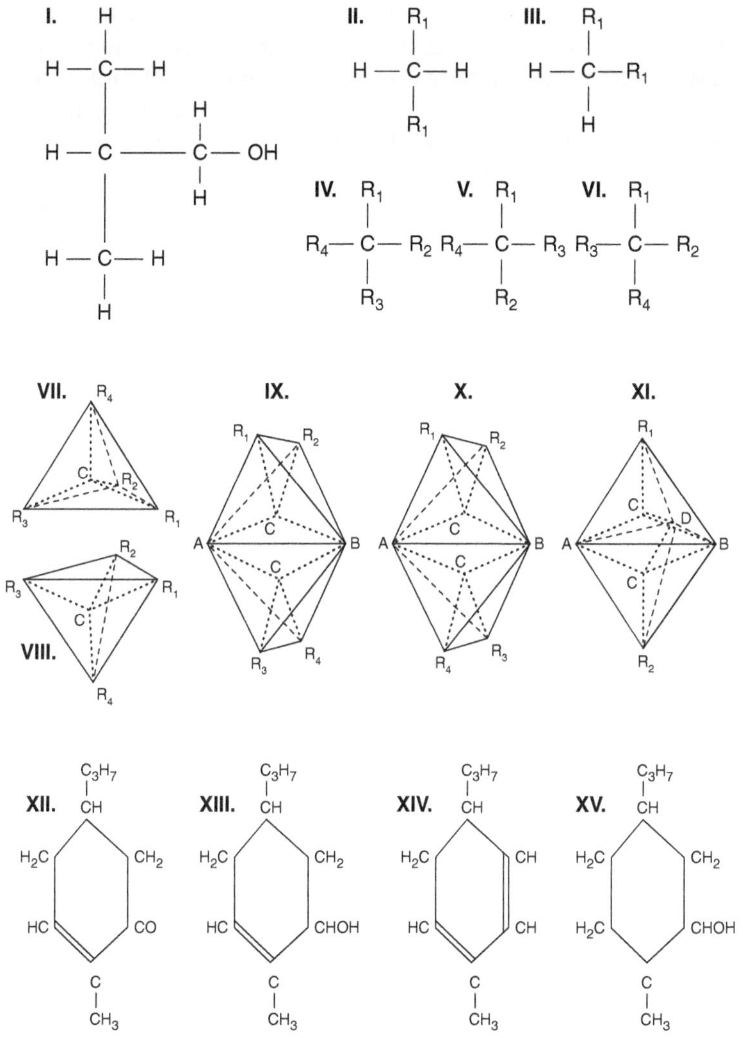

Figure 4.2 Diagrams from van't Hoff's original pamphlet
Source: van't Hoff, 1874/2001.

solutions to an asymmetric arrangement of atoms in the *molecule*. And yet the influence of Pasteur's work on van't Hoff's is less direct than one might think (Ramberg, 2003). In this respect, it is helpful to contrast van't Hoff's work with that of J.A. Le Bel (Snelders, 1975). Just a few months after the publication of van't Hoff's pamphlet, Le Bel also advanced the hypothesis of the tetrahedral carbon atom, independently from van't Hoff (see Le Bel 1874/1963). Le Bel's arguments were much more closely linked to the work of Pasteur and others of the French crystallographic tradition. Le Bel began by considering the symmetry properties of the entire molecule, and arrived at a tetrahedral arrangement for carbon bonded to four different groups as a special case. Unlike van't Hoff, Le Bel did not suggest that *all* carbon atoms were tetrahedral: carbon atoms in methane derivatives might be tetrahedral, while those in more complex molecules are not. In each case, according to Le Bel, we should begin by considering the symmetry of the molecule as a whole.

By contrast, van't Hoff's starts from his hypothesis that all carbon atoms are tetrahedral, and proceeds by building up larger molecules out of these smaller units. As he does so, van't Hoff shows surprisingly little concern with the symmetry properties of the entire molecules. For example, when he turns to consider molecules that contain a double bond, van't Hoff supposes that they are built up from two tetrahedral carbon atoms joined by an edge (see diagrams IX and X in Figure 4.2). And he predicts the existence of isomers in this case, even though the resulting molecules contain a plane of symmetry and so should not be optically active. Similarly, as Peter Ramberg (2003, p. 64) observes, van't Hoff dismisses the hypothesis of a square planar arrangement of atoms not because the resulting molecules would be symmetrical, but because the number of isomers so predicted disagrees with experiment. And at one point, van't Hoff writes that

> if perhaps the asymmetric carbon atom does not make each compound which contains it optically active, it should, in consequence of the basic hypothesis, cause an isomerism that will be apparent *in some way*. (van't Hoff 1874/2001, p. 70, emphasis added)

It is clear throughout his work, then, that van't Hoff's main concern is not with optical activity *per se*, but with the established task of nineteenth-century chemical theory: the explanation and classification of isomerism. As he himself later put it, 'Le Bel's starting point was the researches of Pasteur, mine those of Kekulé' (van't Hoff,

1898/1998, p. 2). Van't Hoff uses the property of optical activity just as chemists used less esoteric properties such as boiling or melting points, as a way of distinguishing between isomers (Ramberg, 2003, p. 64). And it was to be his work, and not Le Bel's, that would be most influential amongst chemists. Van't Hoff's 'proposal for extending the currently employed structural formulae in chemistry into space' promised to allow chemists to explain many more cases of isomerism than was currently possible, including the absolute isomers that had puzzled Wislicenus. And yet, in doing so, van't Hoff also proposed a radical change in the meaning of those formulas: rather than representing Butlerov's chemical structure, van't Hoff's formulas and diagrams purported to show the actual spatial arrangements of atoms within molecules. (In contrast, Le Bel did not offer any perspective diagrams or models to support his reasoning.) In van't Hoff's work, chemical structure now became an aspect of the molecules' spatial arrangement. By making this identification, van't Hoff also crossed epistemic and disciplinary boundaries, expanding the chemist's domain into territory previously thought the preserve of physical studies.

4.1.4 The reception of van't Hoff's work

Van't Hoff's proposal was a bold one. As we have seen, throughout the 1860s, many chemists were unwilling to commit to even the existence of physical atoms. And yet van't Hoff's pamphlet offered speculations regarding the atomic arrangements of numerous organic compounds. The boldness of van't Hoff's work was recognised by its most famous and ardent critic, Hermann Kolbe. As editor of the *Journal für praktische Chemie,* Kolbe wrote an animated attack on the German edition of van't Hoff's work, *Die Lagerung der Atome im Raume* (1877). In an article entitled 'Signs of the Times', Kolbe warned against a return to speculative Naturphilosophie:

> Whoever thinks this worry seems exaggerated should read, if he is capable of it, the recent phantasmagorically frivolous puffery...on 'The Arrangement of Atoms in Space'...A Dr J. H. van't Hoff, of the Veterinary School of Utrecht, finds, it seems, no taste for exact chemical research. He has considered it more convenient to mount Pegasus (apparently loaned by the veterinary school), and to proclaim in his 'La chimie dans l'espace', how, during his bold

flight to the top of the chemical Parnassus, the atoms appeared to him to be arranged in cosmic space...It is typical of these uncritical and anti-critical times that two virtually unknown chemists...pursue and attempt to answer the deepest problems of chemistry which probably will never be resolved (especially the question of the *spatial* arrangement of atoms) and moreover with an assurance and an impudence which literally astounds the true scientist. (Kolbe, 1877, as translated in Rocke, 1993, p. 329)

Perhaps surprisingly, however, Kolbe's line of criticism found little support amongst chemists. In fact, van't Hoff's proposal met with remarkably little opposition. Though he initially received few responses to his proposal, those that van't Hoff did receive were nearly all positive. When his ideas eventually came to enjoy a wider audience, aided by the work of Wislicenus and Adolf von Baeyer in the late 1880s, there remained relatively little controversy over their acceptance.

Of course, criticism was not entirely absent. In an article of 1880, for example, Wilhelm Lossen questioned the theoretical grounding for van't Hoff's representation of double bonds. In van't Hoff's diagrams, single carbon-carbon bonds appeared as straight lines directed between the bound carbon atoms. And yet in their representation of double carbon-carbon bonds, van't Hoff's diagrams seemed to imply that the lines of valence could be bent. Lossen was prompted to ask, in this case, 'what do the touching corners of the tetrahedra mean?' (Lossen, 1880, p. 337, as translated in Ramberg, 2003, p. 93) Lossen did not limit his criticism to van't Hoff's work, however, but also questioned the highly problematic nature of valence itself. What kind of force was chemical affinity? How could this force be 'split' to act only in specific directions? And despite his broad questioning of the theoretical implications of van't Hoff's work, Lossen accepted implicitly its central claim that the valencies of the carbon atom were distributed tetrahedrally.

In this respect, Lossen's criticism is representative of the limited opposition that did exist to van't Hoff's ideas. In general, this opposition was far less wide ranging than Kolbe's attack, and focused largely on specific aspects of van't Hoff's views, such as his treatment of double bonds. By contrast, the core of van't Hoff's proposal, the tetrahedral carbon atom, was received almost without opposition. In his recent history of the development of stereochemistry, Ramberg writes that 'there were essentially no broad debates between chemists about the utility or the reality of the tetrahedron' (2003, p. 330). In fact, the 'reception of the tetrahedral carbon atom could best be described as an application or expansion of van't Hoff's basic ideas to additional compounds and elements, but without any discussion of alternative theories' (ibid., p. 329).[1]

4.2 Building molecules

Why was van't Hoff's proposal so readily accepted by chemists? As we have seen, despite its striking theoretical implications, van't Hoff's pamphlet stayed within the principal remit of much chemical theory, namely the explanation of isomerism. And van't Hoff also retained the chemist's main tools for carrying out this task: structural formulas and, increasingly in the years prior to van't Hoff's pamphlet, three-dimensional, physical models.

4.2.1 Models before van't Hoff

Soon after Crum Brown first introduced his graphical formulas, they would be used as the basis for developing three-dimensional, physical models. At a prestigious Friday Evening Discourse of the Royal Institution in London in 1865, and before a distinguished audience which included the Prince of Wales, August Hofmann demonstrated the promise of the new chemical theory using three-dimensional versions of Crum Brown's formulas (Meinel, 2004). In Hofmann's 'glyptic formulae', the circles and dashes of Crum Brown's formulas were replaced by coloured table croquet balls and connecting rods (see Figure 4.3). By the late 1860s, the use of physical models was increasingly common, and versions of Hofmann's glyptic formulas were soon made commercially available. In 1867, an editorial of *The Laboratory* magazine reported that a set of such models, containing 'seventy balls in all', were available from a 'Mr Blakeman of Gray's Inn Road' ('Glyptic formulae', 1867).

At around the same time, Kekulé himself introduced a different set of models, developed in response to a weakness that he identified in the formulas and models offered by Crum Brown and Hofmann. The problem, Kekulé noted, was that double or triple bonds could not be represented without 'bending' the valence lines linking the atoms

Marsh-Gas   Olefiant Gas   Dutch Liquid

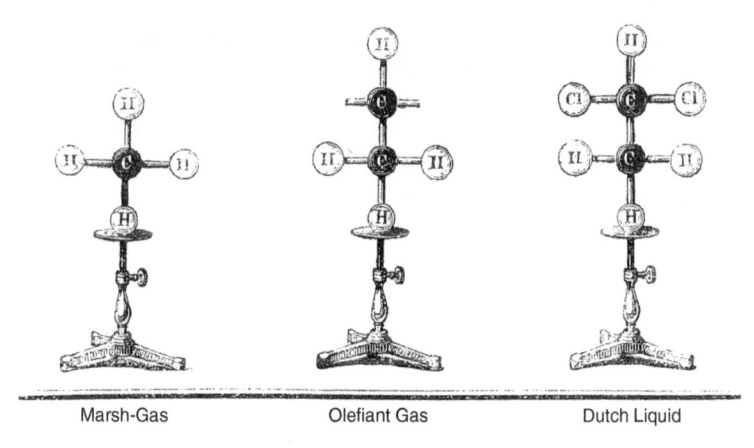

Figure 4.3 Hofmann's 'glyptic formulas'
Source: Hofmann, 1865, p. 189.

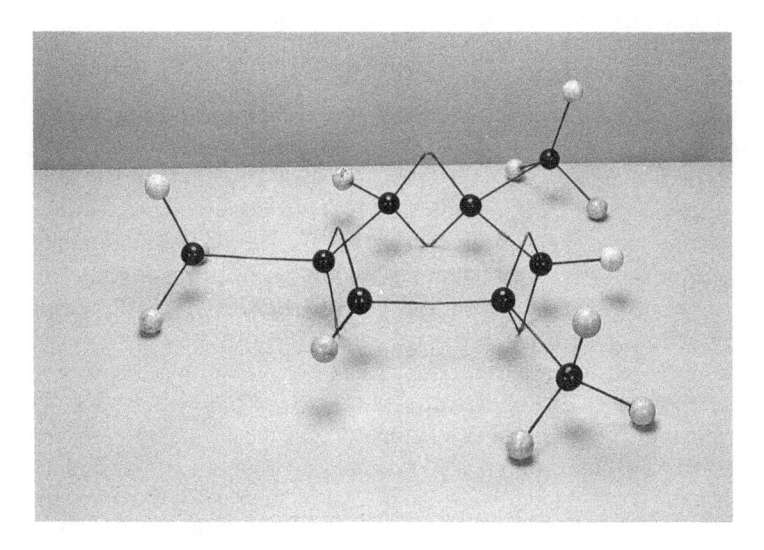

Figure 4.4 Kekulé's tetrahedral models
Source: Museum for the History of Sciences, Ghent University.

(as shown in the formula for vinyl chloride in Figure 4.1). In response to this difficulty, Kekulé proposed instead that the lines of valence be positioned in a tetrahedral distribution (see Figure 4.4). Crucially, however, this tetrahedral orientation was *not* intended to show the spatial arrangement of the bonds. It was merely a practical measure to allow multiple bonds to be represented without bending the wires. Kekulé used these tetrahedral models in his lectures at Ghent, and they were soon put to work as research tools (Meinel, 2004).

4.2.2 Carbon in cardboard

Van't Hoff first mentions his own models in a footnote included in *La Chimie dans l'Espace*, the expanded edition of his work that appeared in French in 1875. Disappointed at the limited response he had received to his pamphlet, van't Hoff sent copies of this refined statement of his proposal to a number of leading chemists. Models were central to this publicity campaign (van der Spek, 2006): the footnote in *La Chimie* not only offers a complete set of the cardboard models to any reader who writes to request them, but also claims that such sets were already in the hands of a number of eminent chemists:

> One may perhaps find some difficulty in following my argument. I have felt this difficulty myself, and I have made use of cardboard models to facilitate the representation. Not wanting to demand too much of the reader, I will gladly send anyone a complete collection of all these objects, such as are already in the possession of MM. Baeyer (Strasbourg), Butlerov (St. Petersburg), Henry (Louvain), Hofmann (Berlin), Kekulé (Bonn), Frankland (London), Wislicenus (Würzbourg), Wurtz, and Berthelot (Paris). For this purpose, write to M. J. H. van't Hoff, Ph.D. Chemist, Rotterdam (Holland). (van't Hoff, 1875, p. 7, as cited in Rocke, 2010, p. 244)

Figure 4.5 shows a set of van't Hoff's models from the Museum Boerhaave in Leiden. Van't Hoff sent these models to his friend Gustav Bremer in 1875, to support Bremer's research into tartaric acid. Van't Hoff himself claimed that his models had helped him to explore the consequences of his ideas, and some authors have even suggested that this may explain some of the developments that occurred in van't Hoff's views between the publication of his original pamphlet and *La Chimie* (Ramsay, 1975; see also van der Spek, 2006).

It seems that a number of chemists wrote to van't Hoff to take him up on his offer of a set of his cardboard models (Ramberg, 2003). After being sent his own copy of *La Chimie*, along with a set of models, Wislicenus greeted van't Hoff's theory with considerable enthusiasm,

Figure 4.5 Van't Hoff's early models of the tetrahedral carbon atom
Source: Museum Boerhaave, Leiden.

and was to play a key role in its further dissemination. Wislicenus not only ordered that the models be reproduced on a larger scale for the use of his students, but also arranged for one of his assistants, Felix Hermann, to produce the (further expanded) German edition of van't Hoff's work, *Die Lagerung der Atome im Raume,* in 1877. Perhaps the most notable addition in *Die Lagerung* was its seven-page appendix, containing templates and detailed instructions, to allow readers to construct their own set of van't Hoff's cardboard models (see Figures 4.6 and 4.7).[2]

Though van't Hoff's models were all tetrahedral, they employed a variety of different methods in order to represent molecules. Although many of van't Hoff's models were regular tetrahedra, some were irregular, and the appendix to *Die Lagerung* also included templates for irregular tetrahedra. Van't Hoff considered the irregular models to be more accurate, since in general the forces between different

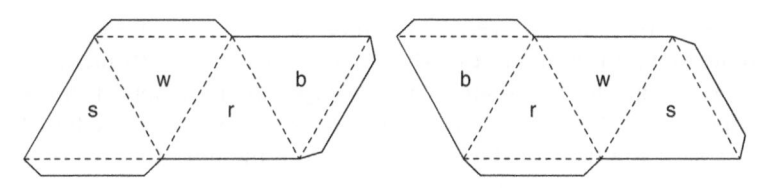

Figure 4.6 Van't Hoff's templates (face bonding models)
Source: van't Hoff, 1877, p. 47.

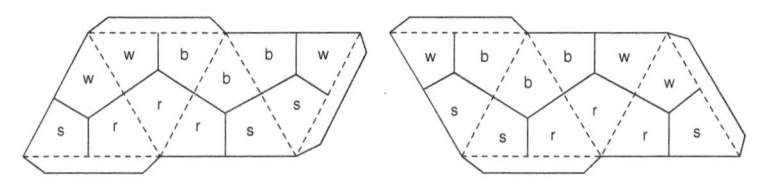

Figure 4.7 Van't Hoff's templates (vertex bonding models)
Source: van't Hoff, 1877, p. 47.

groups in the molecule would be unequal. The regular models were still to be used for most purposes, however:

> If we wish to represent only the two possible formulae [of the asymmetric carbon atom] their peculiar lack of symmetry, their object-and-image relation, and the way they may be rendered identical, the regular tetrahedron with variously coloured corners quite suffices. (1898/1998, p. 8)

Van't Hoff also employed two different methods in order to represent carbon-carbon bonds, which resulted in two different ways of representing molecules as a whole. The first method corresponded most closely to the perspective drawings of his original pamphlet and showed the substituted groups by using letters at the vertices of the tetrahedra. Carbon-carbon bonds were then represented by touching, or sometimes overlapping, vertices of two tetrahedra. In the second method, it was the faces, rather than the vertices, which were coloured to represent the different substituted groups. In these models, valence lines were taken to be directed towards the centre of the tetrahedron's faces, and carbon-carbon bonds were represented by joining the faces rather than the vertices. Both methods were thought to have their own advantages and both appear in the appendix to *Die Lagerung*, in which the reader is sometimes advised to colour the faces of the tetrahedra, and sometimes their vertices (see Figures 4.6 and 4.7).

4.2.3 The role of models

While van't Hoff's work proposed a radical change in chemical theory, his models display a significant continuity with those already used by chemists. Indeed, their similarity to Kekulé's tetrahedral models, constructed seven years prior to the publication of van't Hoff's pamphlet, is remarkable. This similarity was acknowledged by van't Hoff himself. In the English translation of *Die Lagerung*, after describing how his own models could be used to represent the isomerism of the asymmetric carbon atom, van't Hoff commented that 'the Kekulé models, improved by v. Baeyer, and sold by Sendtner (Schillerstrasse 22, München), may be used for the same purpose' (1898/1998, p. 8). By 1885, Adolf von Baeyer had replaced the brass connecting tubes in Kekulé's models with an adjustable joint, and the resulting 'Kekulé-von Baeyer' models were used by chemists well into the 1930s. Prior to the publication of his pamphlet, van't Hoff spent the months between the autumn of 1872 and the summer of 1873 in Kekulé's laboratory in Bonn. Kekulé's tetrahedral models were already in use at Bonn at the time and it is likely that van't Hoff encountered them during his stay there, although there is no direct evidence of this (Meinel, 2004; Ramberg, 2003).

A number of historians have pointed to the growing popularity of graphical formulas and physical models throughout the 1860s and 1870s as an important factor in the development of stereochemistry (Meinel, 2004; Ramberg, 2003; Ramsay, 1981). Thus, Christoph Meinel has recently argued against 'historians of chemistry [who] have treated the emergence of stereochemistry as a sequence of arguments and discoveries within the development of chemical theory' (2004, p. 243). According to this view, 'molecular models have been seen as merely illustrating theoretical concepts such as atom, valency or space' (ibid.). By contrast, Meinel argues that

> the change that eventually resulted in a three-dimensional representation of molecules was led, not by theory, but by modelling – a

kind of modelling invented, not primarily to express chemical theory, but rather as a new way of communicating a variety of messages. (Ibid.)

Chemists' initial move to three-dimensional models, Meinel suggests, was driven by their desire to portray themselves as the architects of the molecular realm, able to construct new substances at will, an image which resonated more widely within the nineteenth-century 'culture of construction'. When van't Hoff published his pamphlet, he was able to build on this existing tradition of three-dimensional modelling (ibid., p. 265).

As we have seen, the importance of models and graphical formulas is clear when we look to the content of van't Hoff's works. Even the title of his original pamphlet explicitly declared its interest in the form of chemists' graphical representations, and van't Hoff began his discussion by considering the square planar arrangement of Hofmann's models and Crum Brown's formulas (cf. Ramberg and Somsen, 2001, p. 67, note 48). Models were also central to van't Hoff's subsequent attempts to develop, explain and promote his ideas.

Of course, while van't Hoff's models were similar to Kekulé's, the latter were originally intended to be representations of chemical structure, not spatial arrangement. As we saw, Crum Brown was also keen to insist that his graphical formulas represented chemical structure, and not the physical position of atoms. Interestingly, however, Crum Brown also acknowledged that his formulas were 'no doubt liable, when not explained, to be mistaken for a representation of the physical position of the atoms' (1864, p. 708). For Kolbe, too, such 'graphical representations' were 'dangerous because they leave too much scope to the imagination' (Kolbe to Frankland, as cited in Rocke, 1993, p. 314). The danger to which Kolbe alluded was also that which had worried Crum Brown, but he invoked a rather higher authority in emphasising its seriousness:

It is impossible, and will ever remain so, to arrive at a clear notion of the spatial arrangement of atoms. We must therefore take care not to think of it in a pictorial way, just as the Bible warns us from making a sensual image of the Godhead. (Ibid.)

Such concerns were widespread, even amongst those who promoted the use of physical models. The editorial in *The Laboratory* thought that, despite the didactic advantages offered by three-dimensional models, '[w]hether they are calculated to induce erroneous conceptions is a question about which much might be said' ('Glyptic formulae', 1867, p. 78). The organic chemist Carl Schorlemmer, who employed physical models in his lectures at Owen's College in Manchester, remarked that 'it happened once that a dunce, when asked to explain the atomic theory, said, "Atoms are square blocks of wood invented by Dr Dalton"' (Schorlemmer, 1894, p. 117, as cited in Meinel, 2004, p. 256).

It was not only Schorlemmer's dunce who strayed from the prescribed interpretation of chemical models, however. In 1869, Emmanuele Paternò, an assistant at Stanislao Cannizzaro's laboratory in Palermo, published a paper on isomerism in halogenated ethanes. At the time, dibromoethane was believed to exist as three isomers, while the structure theory allowed only two. According to Paternò, the three isomers were 'easily explained...when the four valencies of the atom of this element [carbon] are assumed to be arranged in the sense of the four angles of a regular tetrahedron' (Paternò, 1869, as translated in Ramsay, 1981, p. 67). Paternò supported his reasoning by including illustrations of Kekulé's models for these isomers. He took the two models for 1,2-dibromoethane, in which the bromine atoms occupied different relative positions, to represent two different isomers, while on their prescribed interpretation they represented only one. Furthermore, Paternò implicitly assumed that the carbon-carbon bond, like the sticks of Kekulé's models, would be stiff, thus restricting movement between the two configurations (Meinel, 2004; see also Ramsay, 1974).

So it appears that even before van't Hoff, the interpretation of chemical models was not always as carefully circumscribed as their makers had originally intended. Indeed, ambiguities in the prevailing interpretation of molecular models and formulas might even be seen in van't Hoff's pamphlet. We find no acknowledgement of the prescribed interpretation of structural formulas as representations of chemical structure in the pamphlet, but instead we are immediately shown the inadequacy of Crum Brown's formulas if taken as representations of spatial arrangement. At one point, van't Hoff even refers to this as 'the usual means of representation' (1874/2001, p. 68).

4.3 How chemical models represent

It seems, then, that we may better understand the widespread credibility accorded to van't Hoff's proposal if we focus on the models that he offered in its support. As we have seen, van't Hoff's pamphlet engaged directly with an established culture of chemical modelling, and his models were remarkably similar to those already employed within that culture. Van't Hoff also sought to justify his models in the same way that previous models had been justified, by counting isomers.

Ursula Klein (2003) has argued that Berzelian chemical formulas, like H_2O for water and C_2H_6 for ethane, played a similarly important role in the birth of experimental organic chemistry in the late 1820s and 1830s. Rather than expressing existing knowledge, Klein suggests, Berzelian formulas acted as productive 'paper tools' which brought new concepts and practices, most notably the synthesising of new organic compounds through 'substitutions'. Alongside her historical account, Klein offers an analysis of Berzelian formulas as representations, which helps to explain how they were able to play a productive role in the development of organic chemistry (Klein, 2003, Chapter 1). Can we say more about how early chemical models represent? And can this help us to explain their positive role in the development of stereochemistry?

Borrowing a distinction from Peircean semiotics, Ramberg characterises structural formulas and models prior to van't Hoff's work as 'symbolic signs [which] represent objects or concepts by convention' (2003, p. 50). With the advent of stereochemistry, however, chemical models were transformed from conventional symbols to iconic signs, which 'mimic or physically resemble the object' (ibid.). Van't Hoff's work thus brought about a radical transition in the forms of representation used in organic chemistry and 'the crux of this transition in meaning in organic chemistry can be described as a shift from symbolic to iconic formulas' (ibid., p. 325). After van't Hoff,

> the *same* chemical formulas, first developed as a convenient symbolic shorthand or mnemonic device to represent a compound's reactions, had by the end of the century become representations of the molecule as an object. (Ibid.)

While Ramberg is right to point to an important shift in the interpretation of chemical models, the symbolic/iconic distinction may not be the best way to characterise that shift.[3] Klein argues that even Berzelian formulas are iconic in some respects, since the letters in the formulas, like atoms, are discrete units. As a result, Berzelian formulas were not 'completely lacking in imagery' (Klein, 2003, p. 25). Instead, they had a certain 'graphic suggestiveness', conveying a 'building-block image' of chemical compounds (ibid., p. 26). And it seems that we may identify further iconic elements in structural formulas and models before van't Hoff. For example, the number of dashes (or sticks) connected to a given circle (or croquet ball) was the same as the number of bonds of the corresponding chemical atom. On the other hand, conventional elements remained even after van't Hoff's work, as Ramberg himself observes (2003, p. 325). Most obviously, different letters or colours were used to indicate different atoms or groups.

Characterising the shift brought about by van't Hoff in terms of a distinction between symbolic and iconic representations also makes it difficult to understand the apparent continuity in molecular modelling (cf. Klein, 2003, p. 29). Ramberg himself observes that the shift in the meaning of chemical formulas received a 'complete lack of commentary' from chemists (2003, p. 100). But if the symbolic/iconic view were correct then the change brought about by van't Hoff's work would certainly be remarkable: a device originally chosen as an entirely arbitrary symbol was now claimed to be shaped like the molecule. It seems unlikely that such a radical shift would have occasioned such little comment.

The make-believe view allows us to offer a better account of the representational properties of early chemical models. Structural models both before and after van't Hoff's work may be understood as props that prescribe imaginings according to certain rules. Before van't Hoff, these rules concerned chemical atoms and structure. The balls of Hofmann's and Kekulé's models stood for chemical atoms and if two balls were connected by a rod then users were to imagine that the corresponding atoms shared a chemical bond. It was not a rule that, if the balls and sticks had a certain spatial arrangement, then the atoms and bonds also had that spatial arrangement. After van't Hoff, the principles governing the use of models changed. Now, the balls and sticks of Kekulé's models stood for physical atoms and bonds, and users were to imagine that the arrangement of bonds in the molecule followed the arrangement of sticks in the model. In this view, chemical models are not merely arbitrary symbols, even before van't Hoff. Instead, their properties are important. What a model says about chemical structure and arrangement is dependent upon its properties. At the same time, this account allows for the important role played by conventions, both before and after van't Hoff's work: painting a vertex of a model white indicates the presence of a hydrogen atom because this is a principle users take to be in place.

Understanding the models in this way fits better with the apparent continuity in chemical modelling. We need no longer view van't Hoff as proposing a wholesale change in the kind of representation involved in chemical modelling. Instead, he may be understood as expanding the principles of generation that applied to chemical models, so that yet more of their properties (namely, aspects of their spatial arrangement) were understood to be representational. Of course, this is not to deny the significance of van't Hoff's proposal; the point is rather that we are able to situate van't Hoff's work more comfortably within the established practice of modelling.

Like the Phillips machine, models before van't Hoff were not simple scale models. As we have seen, such cases can still be accommodated by the make-believe view. Although chemical atoms were taken to be dimensionless entities, one can still think of Hofmann's and Kekulé's models as asking us to imagine things about the bonds between them, just as water flowing in a pipe of the Phillips machine might mean that we are to imagine that taxes are being paid. In both cases, however, our imagination may run away with us a little, tempting us to think of chemical bonds as physical connections between atoms or money trickling (or gushing) out of people's pockets to the taxman. This is, of course, exactly what worried Kolbe. But it also appears to be what made early chemical models productive forms of representation, pointing to fruitful new ways in which to understand molecules. This process continued even after van't Hoff. In 1885, von Baeyer suggested that the strain that resulted in the wires of Kekulé's models when they were bent to form small rings indicated the strain in the bonds of the corresponding molecules, explaining their tendency to explode.

Meinel suggests that one reason for the success of chemical models was their ambiguity: allowing for a range of interpretations, 'models mediate between audiences without dividing them as theoretical or ideological language would do' (2004, p. 270). Similarly, Klein argues that the success of Berzelian formulas depended partly on their ability to take on different meanings in different contexts, sometimes signifying chemical 'portions', for example, and sometimes submicroscopic atoms. Chemists could therefore use the formulas without committing to a particular foundational theory (Klein, 2003, pp. 14–23). As we have seen, ambiguity in the interpretation of chemical models continued with van't Hoff's work. Van't Hoff did not offer a detailed theoretical interpretation of his models, and many other chemists were content to follow his lead. Others did try to put the models on a firmer theoretical footing, and sometimes interpreted them in a rather different way than van't Hoff himself. For example, while van't Hoff took the models to show the distribution of bonds *around* the carbon atom, Wislicenus seemed to regard the atom *itself* to be tetrahedral (Ramberg, 2003, p. 149). Allowing for such varied readings, van't Hoff's models could be retained while the search for their best interpretation continued.

Such ambiguity and flexibility in the interpretation of chemical models does not pose a problem for the make-believe view. We are used to the idea that viewers of art or readers of fiction may differ in their interpretations of a work. Similarly, van't Hoff's models may have been subject to different principles of generation within different contexts. Just as with art and fiction, the dominant interpretation of a model may also change over time. In Section 4.2.3, we saw that the interpretation of chemical models may have begun to change even prior to van't Hoff's pamphlet. The make-believe view allows that the interpretation of models may change, even without any explicit reinterpretation. Although principles of generation may be explicitly declared, they often remain implicit. A further advantage of the make-believe view is that which we noted in Chapter 3: it is able to accommodate models which do not represent any actual object. This is particularly important in the case of chemical models, since they were used as part of the new synthetic chemistry. In this context, a key function of chemical formulas and models was to represent compounds which did not exist in nature, but which might one day be created in the laboratory (see Ramberg, 2001).

The make-believe view of chemical models is, I think, a very natural one. Meinel links the increasing use of three-dimensional models in chemistry to the growing popularity of children's construction toys that originated with the kindergarten movement (2004, pp. 266–9). In a similar vein, Rocke introduces his readers to the structure theory by asking them to think of a set of Tinkertoys from their childhood (2010, pp. xx–xxi). In fact, Rocke presents his recent history of the development of atomism in nineteenth-century chemistry as an extended argument for the importance of the imagination in scientific thinking. Chemists' ability to imagine the unseen microworld of atoms was a 'pillar' of their methodology (ibid., p. xiii). Rocke contrasts his emphasis on the imagination with other recent studies, which focus on the material form of representations in science, such as Klein's analysis of formulas as 'paper tools'. The make-believe view helps us to see that these two approaches need not conflict, however. In fact, in chemical modelling, the imagination and the material are closely connected: what scientists are to imagine depends upon the material properties of the models they are using.

4.4 Conclusion

Early chemical models offer a good example of the important role that models play in science. In this chapter, we have seen that the make-believe view is able to make sense of these models, allowing us to understand the representational properties of early chemical models and the positive role they played in the development of stereochemistry. Rocke worries that the fleeting and personal nature of the imagination presents a barrier for historians wishing to study it (2010, p. 328). As we saw in Chapter 2, however, fictionality is public. What matters is not what individual scientists imagine, but the shared rules which govern these imaginings. Nevertheless, in order to explore the imaginings engaged in by users of chemical models in more detail, it will be useful to witness the use of these models at first hand. That is what we will do in the next chapter.

5
Playing with Molecules

Chapter 4 examined the role models played in the development of our current three-dimensional conception of molecules. In this chapter, we will consider molecular modelling today. Does the make-believe view provide a good account of the way that molecular models are used, and the attitude that users take towards them? Our assessment will be based on an empirical study examining both hand-held physical models, made out of plastic balls and connecting rods, and computer modelling software. One way to find out if children are engaged in a game of make-believe is to listen to what they say when they are playing the game. If we see a child standing astride a broom shouting 'giddy up!' while his friend complains 'it's my turn to ride now!', we quickly guess that they are pretending that the broom is a horse, that standing astride the broom counts as riding the horse, and so on. Of course, such evidence is not conclusive (the children may be hallucinating). Nevertheless, it provides us with some indication that the children are playing a game, and suggests what form their game takes. Similarly, in this chapter, we will assess the plausibility of the make-believe approach by examining the actions carried out by users of molecular models, and the way that they talk about those actions.[1]

I describe the empirical study and its results in detail in Section 5.1. In Section 5.2, we will find that users' interaction with molecular models suggests that they do imagine the models to be molecules, in much the same way that children imagine a doll to be a baby. As we shall see, however, this captures only part of what is going on in molecular modelling. The children playing with the broom not only imagine things of the broom (that it is a horse), they also imagine things of *themselves* (that they are riding the horse, or stroking it). This imaginative *participation* in games of make-believe is a central and distinctive feature of Walton's theory. In Chapter 2, I suggested that it offers us a way to understand scientists' talk about theoretical modelling. In Section 5.3, we will see that scientists' imaginative participation in modelling extends beyond merely verbal participation, encompassing both their visual and tactile engagement with models. Finally, in Section 5.4, I will suggest that this participation changes our understanding of how scientists learn about the world through scientific modelling. As a result, we are also able to achieve a better appreciation of the value of three-dimensional, physical models over other forms of representation.[2]

5.1 Molecular models in use

5.1.1 The study

One way to try to understand a practice is to listen as it is explained to a newcomer. For this reason, our study involved an experienced user of molecular models (the 'teacher') and three new to the practice (the 'students'). The 'teacher' was a final year doctoral research student in the Department of Biochemistry at the University of Cambridge, who uses molecular models regularly in her research. The 'students' were novices with no education in chemistry or other natural sciences beyond the age of sixteen. The students were set the task of determining the different possible conformations for a number of simple organic molecules, such as propane and cyclohexane, once provided with the correct structural formulas of these compounds. (Different *conformations* of a molecule are different ways in which the atoms within it may be arranged by rotating around, but without breaking, interatomic bonds.) This task was to be carried out using both a ball-and-stick physical model set and a computer modelling program.

The ball-and-stick models used in the study were a Molymod 'Organic Stereochemistry Set'.[3] These models are made available to undergraduate students at a low price by the Chemistry Department at the University of Cambridge, and they are used by students throughout their courses. Indeed, use of such models is recommended in many undergraduate textbooks. At the start of its chapter on stereochemistry, for example, Clayden, Greeves, Warren and Wothers' *Organic Chemistry* advises students that

> In reading this chapter you will have to do a lot of mental manipulation of three-dimensional shapes. Because we can represent these shapes only in two dimensions, we suggest that you make models, using a molecular model kit, of the molecules we talk about. (2001, p. 381)

The model set consists of plastic spheres of different diameters representing standard organic elements, such as carbon, hydrogen and oxygen. Short connecting devices are included so that these spheres may be attached directly to create 'compact models', commonly referred to as 'space-filling' models. Alternatively, connecting rods allow the same spheres to be used to form ball-and-stick models. When used in this way, the models also allow the representation of double bonds between atoms using longer, more flexible connecting rods. It was in this ball-and-stick form that the models were used throughout the study. Used as ball-and-stick models, the Molymod models are broadly similar to Kekulé's tetrahedral models which we discussed in Chapter 4.

The computer modelling software used for the study was the MDL Chime molecular modelling program. This software is available at no cost and it is widely used.[4] It functions as a 'plug-in' for internet browser programs and enables users to view chemical structures in a variety of formats, including space-filling and ball-and-stick formats. In addition to viewing molecular models, users may rotate the entire model on the screen using the computer mouse. The program also allows users to move different parts of the model with respect to others, rotating about parts of the model representing single carbon-carbon bonds, for example. The Chime software requires users to input a database of model files, and for this purpose, the study made use of the 'DCU Molecular Viewing Gallery' of the School of Chemical Sciences at Dublin City University (Pratt, 2006).

The format for the study was as follows. The teacher was first asked to show the student how to carry out the task of determining a molecule's conformations from its structural formula by demonstrating the process through a simple example, typically ethane. The student then attempted the task, with the teacher allowed to prompt and instruct the student when necessary. Both the teacher and the student were asked to reason out loud as far as possible, and the entire study was filmed with a video camera. The recording was then transcribed. The teacher was allowed to select all of the example molecules used for the study. Neither the teacher nor the students were told anything about the study's aims, aside from that it was concerned with how people learn using molecular models. The study was repeated three times with the same teacher and three different students. In the excerpts from the transcripts included below, the teacher is denoted by 'T' and the students by 'S1', 'S2' and 'S3'. Notes in square brackets describe actions undertaken by the teacher and students as they talk, while round brackets show parts of the transcript where their speech was difficult to follow or inaudible. Ellipses indicate pauses.

5.1.2 Talking about models

Perhaps the most striking feature of the teacher's and students' use of molecular models is their tendency to talk as if they were discussing the molecule itself, rather than a model of it. Here are just a few examples:

> T: so what I'm going to do now is take out the double bond [starts disassembling model] so we're back to just four in a row
> S3: hmm
> T: and then I'm going to add some other groups to it
> S3: hmm
> T: so here I've got two carbons, the two black ones
> S3: hmm
> T: and err a red one which is an oxygen and again I just fill up the gaps with hydrogen
>
> ...
>
> T: [starts building model] basically, I'm making just some extra carbons to go on...
>
> ...
>
> S2: OK...hmm...it can go in a variety of...hmm...all the spaces need to be filled up with hydrogens
>
> ...
>
> T: yeah, so basically, when there's just an extra carbon on, you're right, there's loads of different places they can go, here, or here, or here [indicates parts of the model] and because

they're quite small [twists parts of the model] they go anywhere

...

T: [disassembles model] so put the hydrogens back on so it's just a chain of hydrogen and nothing... and... what I'm going to do instead is put on some extra carbons and rather than hydrogen this red one is actually oxygen... and again it needs a hydrogen on it because there's two (holes there) [assembles model] OK, so if you knew that somewhere on our chain of four...

S2: hmm

T: there was an extra carbon with an oxygen and a hydrogen... so these need to go on your chain of four somewhere... what kind of different orientations...

S2: so they can go anywhere where there currently are just a hydrogen... OK, so... well, these could both be coming off the same carbon one, or they could be coming off a different one in any arrangement of that, or obviously off the same one they could both be coming off next to each other or opposite or, erm... [manipulating model]

...

S3: OK, so we're trying to make a molecule, basically

T: yeah

...

T: so what I'm going to ask you to do is to make a similar structure but with four carbons instead of two in it

S3: hmm... so do I have to put the hydrogens on as well?

T: yep... you can start with what I've already made

S3: yep

T: to make it easier rather than starting from scratch

S3: yep, sure, OK, err [picks up physical model] right, so we're going to get four carbons and put them in a row... so do these carbons all have to be all straight in a line or can I put them... like, can I put them here?

In each of these examples, the black spheres of the physical models are typically described as 'carbons', or sometimes 'carbon atoms', the white spheres as 'hydrogens', the connecting rods as 'bonds' and so on. Both the teacher and the students talk about 'taking off' hydrogens or 'adding' a double bond. Constructing the model is described as 'making a molecule'. By contrast, only very rarely did the teacher and student discussions explicitly acknowledge that the immediate objects of their attention were not atoms or molecules, but coloured plastic spheres.

5.1.3 Looking at models

In addition to talking about the model as if it were the molecule, both the teacher and the students repeatedly speak as if they can *see* the molecule. A good example of this is the way that the teacher explains the difference between the 'eclipsed' and 'staggered' conformations of ethane. As the first excerpt shows, this way of speaking was not confined to physical models, but was also used when talking about the computer modelling program.

> T: so if we look down this middle carbon-carbon bond [holds up model to S3's eye line] it's that one there and you can see that these hydrogens are lined up
> S3: hmm
> T: but they can [be] twisted and they can be like that [rotates model]
> S3: aha
> T: so, yeah, different isomers basically, so you can also...so you can see it on the plastic models but you can also see it on the computer [takes mouse]
> S3: aha
> T: so I've got it opened here...I'll try and stop it rotating...so here you can see the hydrogens are all lined up in front of each other
>
> ...
>
> T: basically, the way they're named [picks up physical model] is from when you look down the middle bond what you see...
> S2: oh [takes model] so this one is that [presents one orientation of physical model] basically
> T: yeah, so the first one, the eclipse one, is when you look down the middle and they're both in line
> S2: hmm

5.1.4 Manipulating models

A number of historical and sociological studies of three-dimensional, physical models have stressed the importance of the physical manipulation that they allow. For example, Eric Francoeur notes that, when used in research, molecular models 'are not simply observed. They are submitted to various manipulations, assembled, probed and measured' (2000, p. 78). This is certainly borne out by our own study. Both the teacher and the students almost continually manipulate the models while reasoning through the problems set. The different acts of manipulation carried out on the models may be divided into three rough categories: building the model or taking it apart; twisting different parts of the model around; and rotating the entire model. Once again, it is striking that, while manipulating the models, the teacher and students almost always speak as if they were manipulating the molecule itself.

> T: think about maybe keeping the middle two carbons [gestures a straight line then points to the centre of the model then gestures a twisting motion] straight initially and then seeing what different things you can do
>
> S1: so I guess obviously you can have the carbons facing (inaudible), you know [gestures using model, turning it into a different orientation] and then basically the same down or up. And then obviously the end ones are much more... I can rotate those (inaudible)
>
> ...
>
> T: ... So what you can see here is basically all the bonds can twist round in various different shapes [twists model] but what would happen if we took two of these hydrogens off [takes off white sphere from model] you're losing two hydrogens and instead you can add a double bond [adds two longer connecting rods] so maybe you could look at what difference that makes to the orientations [passes over model to S1]
>
> S1: [picks up model] well, it makes it much more fixed [tries to twist model across double connecting rod, and rotates around end of single connecting rods] so you've got much less movement... basically the ends can twist and that's about it... that's the main thing

T: that's right. With single bonds you can get a lot of movement around them but with double bonds you can't.

...

T: because there's another hole there which is the equivalent of another electron they'll be one (inaudible) [builds model] and do the same on the other side...is there, how to ask it?...how many orientations I guess are there?

S1: well, I guess you can twist these up and around again [twists model] although again that might be problematic (inaudible) [shows that two parts of the model knock into each other] yeah, I mean, they're single things as well, so you can twist the lower half and the upper half around

T: hmm

S1: and you can also twist each branch

T: yeah, branches of the chain

S1: sort of around [twists different parts of the model] and then I guess with these in you can sort of move the hydrogens [twists other parts again]

...

T: erm, right now you're twisting around that middle bond

S3: oh yeah

T: twisting around but when you've got the double bond in

S3: we can't twist it...yeah yeah you can't twist around that way

T: basically double bonds are quite fixed, you can't twist them

...

T: but the hydrogens never kind of hit together did they?

S3: before?

T: like with these extra groups on

S3: no.

Manipulation is also important in the use of computer models. The teacher points out that the computer model will 'let you twist things' and, once again, both the teacher and the students often talk as if they were manipulating the molecule itself:

T: ...Also on this programme it'll let you twist things round...so if you go to this thing called sculpt mode, it lets you just twist things....so you can play about with it, like you can with the model here [picks up physical model] you can do similar things

> T: ... basically if you click on one of the atoms and drag a bit it'll let you play around a bit... so if you drag on a hydrogen you don't get much movement because they can't move much...
>
> ...
>
> S2: hmm... [manipulating computer model using mouse] so this is showing me that I can rotate that and I can rotate that...
>
> T: hmm
>
> S2: and I can't rotate the carbons... is that right, should I be able to rotate the carbons?
>
> T: yeah they should do a bit... that's what we (inaudible) but when you move one of the atoms on the computer what happens to the other ones... do they all stay the same, or...
>
> S2: well, if I move the middle ones the other ones move as well but on the outside ones I can do things quite freely

5.1.5 Play, pretence and realism

One aspect of the teacher and students' discourse that is particularly striking for our purposes are their explicit references to the use of molecular models as 'play'. Frequently, the teacher points out that molecular models allow you to 'play about with' the model or molecule. Molecular models are said to be 'fun' to use, and both teacher and student laugh and joke while using them. At one point, the teacher explicitly refers to pretence:

> T: [building model] so now if we pretend that, rather than just being a carbon like this there's actually an extra... what to put on? maybe add an oxygen onto it. That's the red one [points to model] that's oxygen.

One student uses this language while expressing scepticism over the use of molecular models:

> S3: I mean it's nice to play with but it doesn't seem like it... I don't... I'm more trustworthy of that really [points to structural formula]

This comment must be seen as an exception, however. In general, references to play and pretence do not go hand-in-hand with scepticism over molecular models' representational accuracy. Indeed, the teacher, experienced in their use, most often exhibits a strongly realist attitude. Introducing the main task of the study, for example, she says:

> T: so I'm just going to try and get you to think a bit about how from what you know...if you know what atoms are in a molecule how that can translate into a 3D-model of what the molecule actually looks like

5.2 Molecular models as props

Does the make-believe approach offer a good framework for understanding the way molecular models are used and talked about?

Let us begin with physical models. As we saw, both the teacher and the students routinely talk about the models as if they were molecules. Attaching a black sphere to the model they describe themselves as 'adding a carbon', while removing a white sphere is 'taking off a hydrogen'. Connecting rods are almost always described as 'bonds'. I suggest that the teacher and student imagine the model to be a molecule, just as a child imagines a doll to be a baby and refers to its parts as the baby's 'mouth', 'arms' and 'legs'. These imaginings are not random, but are guided by the model, together with certain rules. The student imagines that each carbon atom in ethane bonds with three hydrogen atoms because the black spheres of the model are connected to three white spheres, and because she understands that if two spheres are connected this means that the corresponding atoms share a bond. And these rules are normative; the student knows that if she were instead to imagine that the carbon was bonded to three oxygen atoms, she would have misinterpreted the model.

Taken together, these observations make molecular models props, in Walton's sense: the models prescribe users' imaginings according to certain rules. Physical molecular models are *reflexive* props: users imagine the models *themselves* to be molecules. The models represent types of molecules, such as ethane, butane or cyclohexane. According to the make-believe view, this is captured by saying that users of molecular models are supposed to imagine that, for example, ethane consists of two carbon atoms, and so on. This may seem the wrong way of expressing the user's attitude to molecular models. Surely, scientists *believe* that ethane contains two carbon atoms.

We need not deny this, of course: one can imagine something that one believes to be true. (Consider the reader of *War and Peace* who imagines that Napoleon invades Russia.) However, we should also remember that some of the content of molecular models is false, and known to be so. The atoms making up the molecule are represented as having definite sizes and locations, for example, although the truth is known to be rather more complicated. It seems that we are merely to imagine that such assumptions hold, not to believe them.

The make-believe view of molecular models receives further support from participants' own descriptions of their use as 'play' or 'pretence'. We have seen that this way of speaking was perfectly compatible with a realist attitude towards molecular models. And, on the analysis I have outlined, the two do not conflict: molecular models can function as props in games of make-believe, while many of the imaginings that they prescribe may be true, and believed to be true. The ethane model asks us to imagine that ethane consists of two carbon atoms, each bonded to three hydrogen atoms, and that these atoms have a certain definite size and structure; the models make these propositions fictional. Some of what the model makes fictional is true (that ethane consists of two carbon atoms) while some is false (that they have a definite location).

What are the principles of generation that govern the use of molecular models? These will vary for different types of models. As we saw in Chapter 4, they have also changed dramatically over time. The principles guiding the use of the models in our study seem to have been as follows. First, there were a number of principles concerned simply with the way that the balls in the model are connected. If two balls are joined by a connecting rod then, fictionally, the corresponding atoms are bonded together. Ball-and-stick models also represent the shape of the molecule. If the connecting rods are distributed in a certain way about an atom, then, fictionally, the bonds between atoms in the molecule are also distributed in that way. Since the spheres of the Molymod models are made to scale, it is also the case that, if one sphere is larger than another, then the corresponding atoms share the same relative sizes. But, of course, not all aspects of the shape of the model are 'carried over' to the molecule. Used as ball-and-stick models, for example, the Molymod molecular models do not represent to scale the distance between bonded atoms, nor, of course, do they represent bonded atoms as being joined by small 'interatomic sticks'. Finally, in addition to representing the bonds between atoms and their spatial distribution, the ball-and-stick models also represent the way in which these bonds allow atoms to move within the molecule: single connecting rods allow for free rotation; double connecting rods do not. Another principle guiding the use of the models, therefore, is that if one part of the molecule is free to rotate in a certain direction, then, fictionally, the corresponding part of the molecule is also free to rotate in that direction.

Let us now turn to the computer modelling program. There are a number of important differences between the physical and computer models. One is that the computer models are not reflexive. Pictures are, in general, not reflexive props. We do not imagine of the framed sheet of canvas entitled *Napoleon Crossing the Saint Bernard* that it is Napoleon. Similarly, users of the computer model do not imagine that the computer display is the molecule that it represents. Nevertheless, I believe, the computer model may still be considered as a prop. The teacher and students imagine things of molecules when viewing the computer model, and those imaginings are guided by the display they see on the screen. Seeing a display depicting two overlapping spheres, for example, users imagine that the corresponding carbon atoms in the ethane molecule share an interatomic bond. Seeing that the two larger spheres can be rotated with respect to each other, the user imagines that the two atoms may also rotate in that direction.

Our empirical study thus provides some support to the make-believe view. Of course, the evidence is not conclusive. There are other possible interpretations of the teacher's and students' actions and talk about molecular models. This is particularly clear in the case of the computer models, in which the analogy to games played with physical objects such as dolls or broom handles is perhaps less strong. Support for the make-believe view is made more compelling, however, when we consider users' active participation in modelling. That is the subject of the next section.

5.3 Playing with molecules

If the discussion so far is correct, then it seems that molecular models may indeed be analysed as props in games of make-believe. If we pay attention only to the models themselves, however, we will miss an important part of what is going on in molecular modelling. To offer a better analysis, we need to recognise the importance of users' active *participation* in modelling. We have already drawn on the idea of participation in Chapter 2, where we saw that we may understand scientists' talk about model-systems as acts of verbal pretence in the game they play with a model. We will now see that there are many other forms of participation, beyond merely verbal participation, and that recognising these forms of participation can help us to understand users' engagement with molecular models.

5.3.1 Participation and depiction

One of the most important aspects of Walton's view is his claim that, when we read a novel or look at a painting, we *participate* in the fiction. As readers or viewers, Walton believes, *we ourselves become props in the games of make-believe we play with these works*. This claim follows closely from the analogy he draws between representations and children's games.

Suppose that David and Anna are playing with a doll. The doll is a prop in their game. David and Anna imagine that the doll is a baby and, depending on what happens to the doll, they imagine various things happening to the baby. If the doll is in the cot, they imagine the baby is in the cot. If doll's 'eyes' are closed, they imagine that the baby is asleep, and so on. Now suppose that Anna raises a cup to the doll. In doing so, she makes it fictional that she feeds a baby. Like the doll, *Anna herself is also a prop in the game*. Anna's actions, together with the principles of the game, prescribe imaginings. Anna is also an object of the imaginings she prescribes. Participants in the game are to imagine *of her* that she is feeding a baby, and *of her act of raising the cup to the doll* that it is feeding a baby. Similarly, according to Walton, we also participate in games we play with novels or paintings, by acting as reflexive props. Reading the early pages of *Dracula*, I am to imagine that I am reading the character Jonathan Harker's journal. It is fictional of me that I am reading Harker's journal. When I look at a painting like *Napoleon Crossing the Saint Bernard*, I am to imagine that I see Napoleon on horseback. It is fictional that I see Napoleon on horseback.

Of course, children's games usually allow for a greater degree of participation than representations like novels or paintings. Children playing with a doll, by moving closer to the doll and picking it up, or putting it down on a pillow, can fictionally move closer to and pick up a baby, or put one to bed. Getting up from your seat in the audience for *Macbeth* and approaching the stage to embrace the lead actor does not count as fictionally approaching or embracing Macbeth. Indeed, neither action would count as fictionally doing anything. Moreover, different representations allow for different forms of participation. Viewing *Napoleon Crossing the Saint Bernard*, we may fictionally see things, notice them or point to them. But if we were to evade the gallery guards and touch the painting this would not count as fictionally touching Napoleon. Touching certain sculptures may count as fictionally touching their subjects, however. Perhaps when a worshipper touches a statue of Christ it is fictional that they touch him. Plays, films and some musical pieces, it seems, allow audiences fictionally to hear things.

Participation underpins Walton's treatment of a number of important problems in aesthetics. As we saw in Section 2.3.3, it plays a key role in his deflationary, pretence account of discourse about fictional characters. In this account, utterances about fictional characters are understood as acts of verbal participation in the game we play with fiction. If we say 'Dracula is a vampire', for example, we are not really asserting something about a fictional entity. We are pretending to make an assertion in the game we play with Stoker's novel. Participation is also central to Walton's analysis of depiction. Consider someone who looks at *Napoleon Crossing the Saint Bernard*. As we saw above, according to Walton, when the viewer looks at the painting she imagines that she sees Napoleon. Moreover, the viewer imagines *seeing* Napoleon. However, this is not sufficient to distinguish pictures from other forms of representation. Someone reading *War and Peace* may also be expected to imagine seeing Napoleon, or St Petersburg. What is distinctive of depiction, Walton thinks, is that the viewer of the painting not only imagines seeing Napoleon, but also imagines *of her looking at the painting* that it is an instance of looking at Napoleon. This is not true of the novel. The reader may imagine seeing Napoleon, but she does not imagine of her reading of the novel that it is seeing Napoleon.[5]

5.3.2 Manipulating molecules

In Chapter 2, I suggested that we understand scientists' discourse about model-systems as acts of pretence in the games they play with a model. Our study of molecular models suggests that scientists' imaginative participation in modelling goes far beyond merely verbal participation, however.

Consider first the teacher's and students' manipulation of the models. We saw the importance of these acts of manipulation in Section 5.1.4. Physical molecular models can be put together, twisted, rotated and finally taken apart and reassembled. Each of these acts is described as if they were carried out on the molecule itself. The teacher and students describe themselves as 'taking off hydrogens', for example, or 'twisting around that middle bond'. I suggest that the way in which users talk about manipulation of molecular models shows that they participate in the games of make-believe they play with molecular models in a variety of ways, just as children participate in their imaginative games.

When Anna raises a cup to the doll, she imagines feeding a baby, and imagines of her act of raising the cup that it is the act of feeding a baby. Her act of raising the cup becomes a reflexive prop in the game. Similarly, I suggest, when the user of a molecular model twists the model, she imagines herself twisting the molecule itself, and she imagines her act of twisting the model to be the act of twisting the molecule. The user's act of twisting the model is a reflexive prop: she is to imagine that it is the act of twisting the molecule, just as the children are to imagine that raising a cup to the doll is feeding a baby. It is fictional that the user's manipulation of the model is manipulation of the molecule.

Precisely what form of participation molecular models allow is a matter of what actions users can perform vis-à-vis the models such that they are imagined to be actions vis-à-vis the molecule. In the case of the ball-and-stick models used in the study, users of the models can fictionally build or disassemble molecules and fictionally twist different parts of them around. They can also fictionally rotate the molecule, and determine that two of its atoms will hinder each other. In so-called 'open' or 'skeletal' molecular models, on the other hand, the distances between atoms in the molecule are represented to scale and users can measure the distance between different parts of the model. And it seems that these acts of measurement are also (at least sometimes) construed as if they were carried out on the molecule. Francoeur quotes English chemist Derek Barton, who remarks of some skeletal models that

> [t]he accuracy of manufacture and scale of these models is such that quite satisfactory measurements with a metre rule can be made of the distance between atomic centres. (Barton, 1956, p. 1137, as cited in Francoeur, 2000, p. 75)

What about computer modelling? As we saw in Section 5.1.4, computer modelling also allows users to manipulate the model displayed on the screen. Just as with physical models, when users perform these 'virtual' manipulations, they appear to imagine them being performed on the molecule itself. While using the mouse to rotate parts of the model, for example, one student asks 'should I be able to rotate the carbons?' Of course, users' virtual manipulation using the mouse is very different from the physical acts of manipulation allowed by physical models. We will consider the importance of this in Section 5.3.4.

5.3.3 Seeing molecules

Earlier we saw that, when looking at the models, the teacher and students speak as if they are looking at the molecule itself. Demonstrating the computer model, for example, the teacher remarks, 'here you can see the hydrogens are all lined up in front of each other'. Moreover, it seems clear that the teacher and student do not simply imagine seeing the molecule, as they might while reading a textbook. They also imagine of their act of looking at the model that it is looking at the molecule. The teacher directs the students to 'look down this middle carbon-carbon bond' in order to distinguish between staggered and eclipsed forms of ethane and points at the relevant part of the model to direct their attention there. Such actions suggest that users imagine of their visual actions vis-à-vis the model that they are visual actions vis-à-vis the molecule.

It seems, then, that molecular models are depictions, in Walton's sense. The teacher and student relate to the model in a similar manner to that in which the viewer of *Napoleon Crossing the Saint Bernard* relates to the painting. Of course, although the teacher and student speak as if they were looking at the molecule when viewing the model, they are well aware of the distinction between the two. Similarly, the viewer of the painting is well aware that she is looking at a piece of canvas, even while she imagines seeing Napoleon. Many features of the model are not 'carried over' to the molecule. We don't fictionally see carbon atoms as black and hydrogen as white. But the same is true of many depictions. We don't fictionally see the subjects of line drawings as composed of black and white lines. Compared with the painting, molecular models are relatively lacking in what Walton terms *richness*. Richness consists in the variety of visual actions that, by virtue of actual visual actions we perform regarding the work, we may fictionally perform. Looking at the painting, we may first fictionally scan the horizon, before peering closer, fictionally examining the red of Napoleon's cloak, or the texture of the horse's mane. The visual games allowed by molecular models are not particularly rich. We can fictionally see the molecule as having a certain structure, but not, of course, its colour or the texture of its surface.

Francoeur points to an important objection that might be made to the claim that molecular models are depictive:

> Molecular models can be considered a mode of visual representation, in the sense that they allow us to visualize molecular structure. Yet, arguably, no chemist would propose that models, even in their more elaborate forms, are about what molecules 'really' look like. In fact, any argument to that effect is bound to be considered technically moot, since, as a prominent biophysicist has noted, 'for something smaller than the wavelength of light, there is no such thing as showing how it really looks on the molecular level'. (Francoeur, 1997, p. 12, quote from Richardson et al., 1992, p. 1186)

The make-believe account helps to alleviate the sense of paradox somewhat. In a resemblance account of depiction, for example, it would be hard to see how molecular models could count as depictive at all. According to resemblance accounts, pictures represent their objects because they look like them. How could molecular models look like molecules if molecules have no appearance? It seems rather less problematic to say that we *imagine* ourselves looking at molecules.

Things are not so straightforward, however. Many molecular models are built to scale. They would therefore seem to represent molecules as being smaller than the wavelength of light. On the make-believe account, this means that we are to imagine that the molecules are smaller than the wavelength of light. But now it appears that we are asked to imagine a contradiction: we are both to imagine seeing molecules and to imagine that they cannot be seen. Similar problems arise for many pictures. When we look at the panels of the Sistine Chapel, for example, it seems we are to imagine seeing the Creation. And yet we are also to imagine that the Creation is unseen. There are several different solutions we might adopt to such problems. For our purposes, we may simply note that the presence of a contradiction like this need not lead us to deny that molecular models are depictions.

5.3.4 Touching molecules

In contrast to our normal use of the term, Walton applies his definition of depiction to senses other than sight. Thus, a musical work that represents the chirping of birds may be depictive with respect to hearing. When we hear the music, it is fictional that we hear birdsong, and it is fictional of our listening to the music that it is listening to birdsong. And a teddy bear may be depictive with regard to touch. It is fictional of our touching the teddy bear that it is touching a bear. I believe that, as well as being depictive with respect to sight, the physical molecular models used in our empirical study are also depictive with respect to touch.

To see this, suppose the scientist tries, and fails, to twist the model in a certain direction. When she does so, she feels the model exert pressure against her hand. In Section 5.3.2, we saw that scientists manipulating molecular models imagine themselves manipulating molecules. In this case, then, the scientist imagines her action to be that of trying and failing to twist the molecule. It follows, I suggest, that the scientist also imagines the pressure she feels against her hand to be exerted by the molecule. A comparison with dolls is helpful here. Suppose a child goes through the action of 'feeding' her doll and finds that its 'arms' get in the way. If she imagines her action to be that of feeding a baby, then it seems that she imagines the pressure exerted on her hands by the doll's 'arms' as they impede her to be exerted by the baby's arms.[6]

Physical molecular models are thus depictive with respect to touch. Once again, however, the games allowed by molecular models are not particularly rich. A teddy bear allows for a rich variety of different tactile actions, each of which may be imagined to be tactile actions involving the bear. We can fictionally squeeze the bear tightly or softly, feel the texture of his fur, and so on. Feeling the model resist twisting in a certain direction counts as fictionally feeling the molecule resist twisting in that direction. But feeling the texture of the balls and sticks that make up the model does not count as fictionally feeling the texture of the molecules. Nevertheless, the depictive character of scientists' tactile interaction with physical models is important, as we shall see in the next section. It is also an important way in which they differ from computer models.

Computer molecular models are not depictive with respect to touch. As we have seen, computer models allow for various forms of participation. Users can manipulate the model displayed on the screen, and when they do so, they imagine manipulating the molecule itself. But, of course, the scientist using a computer model does not manipulate the model with her bare hands. She does so by moving the mouse and witnessing its effect on the screen. It is this 'virtual' manipulation that the scientist imagines being carried out on the molecule, not the actual, physical motion her hand makes with the mouse. And the tactile element is therefore lost. When she moves the mouse to perform this virtual manipulation, the scientist does, of course, feel the mouse press against her hand. And an accomplished user of computer models may even imagine feeling the molecule resist as she does so. But she does not imagine of the steady, even, feel of the mouse in her hand that it is the feeling of a molecule pushing back against her. For this reason, computer models (or at least those used in this study) are not depictive with respect to touch.[7]

5.4 Imagined experiments

To sum up so far: we have seen that molecular models may be analysed as props in games of make-believe, and that users of molecular models participate in these games. Users of molecular models not only imagine things about molecules, they also imagine themselves looking at molecules, picking them up and twisting them around. If this is correct, and scientists do engage in these sorts of self-imaginings, then clearly it is an advantage of the make-believe account that it is able to capture this feature of the use of molecular models. And users' participation in modelling lends further support to the connection the account draws to games of make-believe. Still, one might wonder whether this participation is important. Is scientists' imaginative participation in modelling any more than an interesting psychological curiosity? I believe that scientists' participation in modelling is important. In particular, I think, it suggests a different view of how scientists use molecular models to find out about the world, through what I shall call *imagined experiments*.

As we noted in Chapter 3, we commonly think of learning about the world through scientific modelling as taking place in an indirect, two-stage, manner. First, scientists investigate the properties of the model. Second, they then use similarities between the model and the system being modelled to translate what has been learned about the model into information about the system. This might be a good description of the way that some physical models are used. For example, consider an engineer using a scale model to carry out tests. In this case, the engineer may well first determine the properties of the model, and then go on to calculate how these results 'scale-up' for the full system. Recognising scientists' participation in modelling, however, suggests that physical models may sometimes be used to find out about the world in ways not captured by the two-stage view.

To see this, let us again consider children playing with a doll. It is often noted that children's games of make-believe play an important role in teaching them about the world.[8] To return to our previous example, suppose Anna raises a cup to the doll's 'mouth'. When she does so, Anna imagines her action of raising the cup to be that of feeding a baby. Intuitively, it seems that Anna can learn something about caring for a baby through actions such as these. She may, for example, learn (at least partly) *how* to feed a baby. And here it is clearly important that Anna does carry out an action, rather than simply saying to herself 'imagine feeding a baby is like this...' Doing so, it seems, further enhances Anna's experience, since she rehearses an action (at least partly) close to that of feeding a baby. Moreover, it would appear that, through these imagined actions, Anna may learn things *about* babies. If Anna raises the cup and the doll's 'arms' get in the way, she imagines that the baby's arms have got in the way of her feeding it, and so learns that babies' arms can get in the way when you feed them.

Notice how poorly a two-stage account would characterise Anna's learning process. Similarities between the properties of the doll and the baby are, of course, important for underpinning her learning. If the doll had three arms, it would seriously lead her astray. But Anna does not first attend only to the doll's properties, and then go on to 'translate' this into knowledge about babies. Instead, her interaction with the doll is itself shot through with her imaginings about the baby. Anna imagines the actions that she carries out with the doll to be actions performed with a baby, and it is through these imagined actions that she learns both how to feed a baby and that babies' arms can get in the way when you feed them.

I suggest that molecular models sometimes teach scientists about the world in a similar way. Suppose that a scientist tries, and fails, to twist the model in a certain direction, and thereby learns that the molecule is rigid in that direction. In doing so, I think, the scientist does not first discover that the model is rigid and then infer that, since the model and molecule share the relevant properties, the molecule is rigid also. Instead, she imagines her actual manipulation of the model to be the manipulation of the molecule, and it is these imagined actions that teach her about the molecule's properties. The scientist imagines the twisting action she performs on the model to be a twisting action performed on the molecule. And imagining trying, and failing, to twist the molecule allows her to discover that, fictionally, the molecule is rigid in that direction. If the imagined actions that the scientist performs are accurate – that is, if what she imagines trying to twist the molecule at that point to be like is close to what trying to twist the molecule at that point would actually be like – then she is able to learn about the molecule's actual properties. Unlike the child playing with the doll, of course, the scientist doesn't want to learn *how* to twist the molecule, because we can't twist molecules with our bare hands. But this form of learning is important with other physical models, such as the silicone models used by doctors to teach breast self-examination (see Francoeur, 1997).

Recent historical studies of a wide variety of physical models in the sciences have stressed the importance of the bodily, tactile engagement that they allow.[9] In his study of the history of molecular models in research, for example, Francoeur observes that 'the way models mechanically resist or yield when one tries to have them adopt some configuration constitutes a physical, embodied experience of "allowed" or "non-allowed" spatial configurations' (1997, p. 16). The value of this tactile element in molecular modelling is stressed by chemists themselves, and it is often cited as the main advantage of physical models over computer modelling programs. Thus, in her discussion of molecular models for a Biophysical Society National Lecture, biochemist Jane Richardson (quoted above by Francoeur) notes that 'the real virtue of ["space-filling" physical models] is the physical "feel" for the bumps, constraints, and degrees of freedom one obtains by manipulating them' (Richardson et al., 1992, p. 1186, cited in Francoeur, 1997).

Recognising the role of scientists' imagined experiments with models allows us to appreciate the importance of the tactile engagement that physical models allow. In Section 5.3.4, we saw that physical molecular models are depictive with respect to touch. When users manipulate physical models, they experience sensations of touch which they imagine to be caused by manipulation of the molecule. This allows scientists to investigate the properties of molecules in a kind of 'imagined analogue' of the way in which we discover the properties of normal, everyday objects. We might find out whether a piece of metal is rigid by trying to bend it and feeling it resist. Similarly, using physical models, scientists can find out whether a molecule is rigid by carrying out imagined actions, fictionally bending or twisting the molecule and feeling the resistance that it exerts. Computer models do not allow us to utilise our sense of touch in this way. For that reason, physical models even now retain a crucial advantage over their computer counterparts, despite the considerable advantages that computer modelling has introduced.[10]

5.5 Conclusion

In this chapter, we have seen that the make-believe view gains support when we look at the way that models are used. Users' interaction with molecular models, and the way that they talk about what they are doing, suggests that they do treat these models as props that guide their imaginings in various ways. As we have seen, this not only includes imagining the balls and sticks of the models to be atoms and bonds. Users of molecular models also imagine themselves looking at molecules, picking them up and twisting them around. These imagined actions are central to the way that such models teach us about the world, and help to explain the value of the tactile, manipulable properties of physical models in particular.

Conclusion

At the start of this book, we set out to answer two questions: what are scientific models, and how do they represent the world? I suggested that we look for answers from an unlikely source, in children's imaginative games.

In Chapter 1, we saw that theoretical modelling poses ontological problems, and that most philosophers respond to these problems by adopting a version of what I called the *indirect view* of modelling. In this view, scientists' assumptions and equations give rise to model-systems, which are some form of abstract or fictional entity, and scientists represent the world via these entities. One of the main aims of this book has been to offer an alternative to this way of looking at scientific modelling. Chapter 1 also looked more closely at the problem of scientific representation. We saw that Callender and Cohen are wrong: even if models represent in virtue of some other more fundamental form of representation, the problem of scientific representation is still one that must be faced.

I have offered answers to both of the questions with which we began, based on the idea that models function as props in games of make-believe. Chapter 2 focused on ontology. We saw that the make-believe view is able to make sense of theoretical modelling without positing any abstract or fictional entities which satisfy scientists' modelling assumptions. In place of the indirect view, I offered a *direct view* of modelling: when scientists model a system, they ask us to imagine things about it. Learning about a model does not involve discovering facts about any abstract or fictional entity. Instead, we learn about a model by exploring the intricate web of imaginings which the model prescribes. When scientists appear to talk about theoretical models as objects, I suggested, we should not take this talk too seriously. Instead, they are pretending, 'going along with' the model in order to show us how it works.

In Chapter 3, we turned to the problem of scientific representation. I argued that the make-believe view offers us an account of how models represent the world, which meets the requirements set out in Chapter 1. According to this account, models represent in virtue of prescribing imaginings. Understanding representation in modelling in this way allowed us to avoid the problems faced by stipulation and similarity accounts. It also provided us with a framework for understanding realism in modelling, since in the make-believe view, both theoretical and physical models prescribe propositions about their objects. Finally, we saw that, unlike existing accounts of scientific representation, the make-believe view is able to explain how it is that models can be representational without representing any real system.

One of the advantages of the make-believe view, I think, is that it fits well with the way that models are used in science, and the attitude that scientists take towards them. In the final two chapters of this book, I tried to show that the make-believe view can help us to understand scientific practice by looking at an important group of models in some detail.

Chapter 4 focused on early chemical models, and in particular, on J. H. van't Hoff's cardboard models of the tetrahedral carbon atom. This case offered an example of the important role that models play in science. Focusing on chemists' models allowed us to reach a better understanding of how we came to our current conception of molecules as three-dimensional, spatial arrangements of atoms. We also saw that the make-believe view is able to make sense of early chemical models, offering an account of these models' representational properties which helps to explain their positive role in the development of stereochemistry.

In Chapter 5, we saw that the view is also supported when we look at the way that chemical models are used and talked about. Users imagine chemical models to be molecules just as a child might imagine a doll to be a baby. And they also *participate* in the imaginative games they play with these models in ways that go beyond the acts of verbal pretence discussed in Chapter 2. Scientists using molecular models, I suggested, perform *imagined experiments* on molecules, imagining themselves looking at molecules, putting them together and breaking them apart.

By now, I hope, games of make-believe do not seem such an unlikely source of inspiration for understanding scientific models. Instead, I think, they offer us a natural and intuitively appealing way to understand models, one which also turns out to be remarkably fruitful in a number of different directions. Much remains to be done, of course. Not least, there remains the task of spelling out the different principles of generation that govern different kinds of models, and of examining the various forms of participation that models allow. But I hope at least to have shown that the make-believe view offers a promising framework with which to understand scientific models, and to have pointed to some of the insights that this framework can provide.

Notes

1 Models and Representation

1. Giere's account is often described as a version of the 'semantic view' of theories. Other versions of the semantic view take models to be set-theoretical structures (Suppes, 1960) or trajectories through phase-space (van Fraassen, 1980). These accounts aim to give a formal analysis of the structure of theories and the notions of model they employ are rather different from those we are concerned with here. For helpful discussions of this issue, see Godfrey-Smith (2006) and Thomson-Jones (2007, 2010).
2. See also Godfrey-Smith (2006) and Hughes (1997).
3. The terminology I use here is similar to that in Friend (2007).
4. Variations of this point may be found in Frigg (2006), Hughes (1997) and Suárez (2003), although each draws rather different lessons from it.
5. This problem is also raised by Suárez (2003) and Callender and Cohen (2006).
6. Of course, this is not to claim that realism in modelling is the same as realism in painting.
7. Notice that the question is not whether the salt shaker could be a model-representation under *any* circumstances, but whether it is a model-representation in the particular circumstances that Callender and Cohen describe, in which we simply stipulate that it represents Madagascar. As we will see in Chapter 3, in other circumstances the salt shaker might be used as a model-representation. But doing so requires more than the act of stipulation described by Callender and Cohen.

2 What Models Are

1. The suggestion that Walton's theory may be applied in the context of scientific modelling has also been made by Anouk Barberousse and Pascal Ludwig (2000, 2009) and Roman Frigg (2010a, 2010b). I consider Frigg's account in detail in Section 2.4.2. The issues addressed by Barberousse and Ludwig are rather different from those that concern us here.
2. This definition is intended to distinguish the world of the representation from that of the games individuals play with it. When we read a novel, Walton claims, we not only imagine the events it narrates, we often imagine that we ourselves hear or read about those events. But we are not a character in the novel (1990, p. 60). I will say more about participation in games with scientific models in Chapter 5.

3. Of course, Walton's analysis of discourse about fictional characters is far from universally accepted. Friend (2007) gives a very helpful review of this debate.
4. In fact, Walton's theory does not demand that the speaker actually engage in pretence. When we say 'St Paul's is damaged' we might specify the relevant kind of pretence without exemplifying it (Walton, 1990, p. 404).
5. And, in fact, Frigg's analysis of model-systems differs from Walton's analysis of fiction at a number of points, and sometimes seems at odds with antirealism. For example, he writes that 'the attribution of certain concrete properties to models... is explained as it being fictional that the model-system possess these properties' (2010b, p. 116; see also 2010a, p. 261). To say that it is fictional that the model-system possesses certain properties, however, is to say that we are to imagine that the model-system possesses those properties. This would seem to conflict with antirealism: we cannot imagine things about model-systems if there are none.
6. My arguments in this section are similar to those in Thomson-Jones (2007, section 4).

3 How Models Represent

1. In stressing the role of propositions, the make-believe account is similar to that offered by Daniela Bailer-Jones (2003). Bailer-Jones describes the relationship between models and propositions as one of nonlogical 'entailment' (2003, p. 60). Critics have objected that this notion of entailment remains obscure (e.g., Callender and Cohen, 2006, p. 70). The make-believe account explains the link between models and propositions: models, together with principles of generation, make propositions fictional.
2. Notice that I do not deny that scientists sometimes talk about the accuracy of models in terms of similarities between model-systems and the world, or that it may be useful to do so. In my view, however, such talk should not be taken literally, but instead understood according to the interpretation I offered in Chapter 2 (Section 2.3.3.2).
3. Note that *Sim* is not endorsed by Giere (2004, 2010), who instead stresses the role played by the scientists using a model. In this respect, the view expressed in this section is similar to Giere's. I offer a critique of Giere's account in Toon (forthcoming).
4. I am grateful to Martin Thomson-Jones for encouraging me to stress this point more clearly.
5. Of course, one response to this might be to adopt a version of the indirect fictions view of theoretical modelling considered in Section 1.2.3. I say why I do not think this would be a good route to take in Section 2.4.
6. 'D.D.I.' stands for 'denotation, demonstration and interpretation'. According to Hughes, these combine in the following way: 'Elements of the physical world are denoted by elements of the model; the model

possesses an internal dynamic that allows us to demonstrate theoretical conclusions; these in turn need to be interpreted if we are to make predictions' (Hughes, 1997, p. S325).
7. The phrase 'deferral strategy' is taken from Thomson-Jones (2007).
8. Callender and Cohen also attempt to defer the problem posed by models without actual objects, observing that 'the worry arises for all species of representation – not just scientific representation – and there is no reason to suspect that whatever ultimately explains representations of unicorns and golden mountains won't work for representation of phlogiston and the ether' (2006, pp. 80–1). There is an important difference between Callender and Cohen's deferral strategy and my own, however. Callender and Cohen simply express a hope that a solution to the problem for other forms of representation may be applied to scientific models. They tentatively suggest a 'Humean strategy', which provides a relational theory for 'atomic' representations and explains representations without actual objects as constructed as 'compounds' of other representations. But they do not show whether this can be applied to scientific models, or whether their account would remain intact if it were. This amounts simply to deferring the problem for scientific representation itself. By contrast, the make-believe account *reduces* the problem for scientific representation to the more general problem for imagination.

4 Carbon in Cardboard

1. It is worth noting that Ramberg's analysis focuses on chemists in Germany (see Ramberg, 2003, p. 8).
2. Van't Hoff's templates might thus be seen as an ingenious attempt to circumvent the 'Latourian' problem of the immobility of three-dimensional models compared to flat inscriptions (see, for example, Latour, 1990).
3. The notion of iconic representations is, of course, put under considerable pressure by Goodman (1976).

5 Playing with Molecules

1. In looking to practice to inform our understanding of models, my study therefore follows work by Daniela Bailer-Jones (2002, 2009), although my methodology is rather different. While Bailer-Jones's study used interviews to determine scientists' views on various aspects of modelling, mine was based on observations of the way that models were used and, in particular, of how this practice was explained to novices.
2. For three-dimensional models in general, see de Chadarevian and Hopwood (2004). Laszlo (2000) offers a chemist's perspective on the relationship between molecular models and toys.
3. These model sets are manufactured by Spiring Enterprises Ltd., Gillmans Industrial Estate, Natts Lane, Billingshurst, West Sussex, England.

4. MDL Chime is a product of Accelrys, Inc., 10188 Telesis Court, Suite 100, San Diego, CA 92121, USA. It is available at http://www.symyx.com/downloads/downloadable/.
5. For criticism of Walton's account of depiction see, for example, Schier (1986).
6. Of course, there are many puzzles one could raise in this regard, just as there are regarding the visual depiction of molecules. For example, isn't it fictional that we cannot touch molecules? Once again, it seems that the threat of contradiction does not prevent us from engaging in such imaginings.
7. However, there are some molecular modelling computer programs being developed that allow for tactile engagement. See Francoeur and Segal (2004).
8. For make-believe and pretence in children's games see, for example, Leslie (1987), Singer and Singer (1990), Sheikh and Shaffer (1979), Weisberg and Bloom (2009) and Wyman et al. (2009).
9. See, in particular, the papers collected in de Chadarevian and Hopwood (2004).
10. This includes, for example, the relative ease with which computer models allow the user to build models of large molecules and their ability to simulate more complex interactions between atoms within a molecule.

References

Bailer-Jones, D.M. (2002). Scientists' thoughts on scientific models. *Perspectives on Science*, *10*(3), 275–301.

Bailer-Jones, D.M. (2003). When scientific models represent. *International Studies in the Philosophy of Science*, *17*, 59–74.

Bailer-Jones, D.M. (2009). *Scientific Models in Philosophy of Science*. Pittsburgh: University of Pittsburgh Press.

Barberousse, A. and Ludwig, P. (2000). Les modèles comme fictions. *Philosophie*, 68, 16–43.

Barberousse, A. and Ludwig, P. (2009). Models as fictions. In M. Suárez (ed.), *Fictions in Science: Philosophical Essays on Modeling and Idealization* (pp. 56–73). London: Routledge.

Barton, D.H.R. (1956). Molecular models for conformationnal analysis. *Chemistry and Industry*, 1136–7.

Butlerov, A. (1861). Einiges über die chemische Structur der Körper. *Zeitschrift für Chemie*, *4*, 549–60.

Callender, C. and Cohen, J. (2006). There is no special problem about scientific representation. *Theoria*, *55*, 67–85.

Cartwright, N. (1983). *How the Laws of Physics Lie*. Oxford: Clarendon.

Cartwright, N. (1999). *The Dappled World: A Study in the Boundaries of Science*. Cambridge: Cambridge University Press.

Chakravartty, A. (2010). Informational versus functional theories of scientific representation. *Synthese*, *172*(2), 197–213.

Clayden, J., Greeves, N., Warren, S. and Wothers, P. (2001). *Organic Chemistry*. Oxford: Oxford University Press.

Contessa, G. (2007). Scientific representation, interpretation and surrogative reasoning. *Philosophy of Science*, *74*, 48–68.

Contessa, G. (2010). Scientific models as fictional objects. *Synthese*, *172*(2), 215–29.

Crum Brown, A. (1864). On the theory of isomeric compounds. *Transactions of the Royal Society of Edinburgh*, *23*, 707–19.

de Chadarevian, S. (2002). *Designs for Life: Molecular Biology after World War II*. Cambridge: Cambridge University Press.

de Chadarevian, S. (2004). Models and the making of molecular biology. In S. de Chadarevian and N. Hopwood (eds), *Models: The Third Dimension of Science* (pp. 339–68). Stanford, California: Stanford University Press.

de Chadarevian, S. and Hopwood, N. (eds) (2004). *Models: The Third Dimension of Science*. Stanford: Stanford University Press.

Downes, S. (2009). Models, pictures and unified accounts of representation: lessons from aesthetics for philosophy of science. *Perspectives on Science*, *17*, 417–28.

Elgin, C. (2009). Exemplification, idealization, and scientific understanding. In M. Suárez (ed.), *Fictions in Science: Philosophical Essays on Modeling and Idealization* (pp. 77–90). New York: Routledge.

Elgin, C. (2010). Keeping things in perspective. *Philosophical Studies, 150*(3), 439–47.

Fine, A. (1993). Fictionalism. *Midwest Studies in Philosophy, XVIII*, 1–18.

Fisher, N. (1975). Wislicenus and lactic acid: the chemical background to van't Hoff's hypothesis. In O.B. Ramsay (ed.), *van't Hoff-Le Bel Centennial* (pp. 33–54). Washington, D.C.: American Chemical Society.

Francoeur, E. (1997). The forgotten tool: the design and use of molecular models. *Social Studies of Science, 27*, 7–40.

Francoeur, E. (2000). Beyond de-materialization and inscription: does the materiality of molecular models really matter? *Hyle, 6*(1), 63–84.

Francoeur, E. and Segal, J. (2004). From model kits to interactive computer graphics. In S. de Chadarevian and N. Hopwood (eds), *Models: The Third Dimension of Science* (pp. 402–29). Stanford, California: Stanford University Press.

French, S. (2003). A model-theoretic account of representation (Or, I don't know much about art...but I know it involves isomorphism). *Philosophy of Science, 70*, 1472–83.

Friend, S. (2007). Fictional characters. *Philosophy Compass, 2*, 141–156. doi:10.1111/j.1747-9991.2007.00059.x.

Frigg, R. (2006). Scientific representation and the semantic view of theories. *Theoria, 55*, 49–65.

Frigg, R. (2010a). Models and fiction. *Synthese, 172*(2), 251–68.

Frigg, R. (2010b). Fiction and scientific representation. In R. Frigg and M. Hunter (eds) *Beyond Mimesis and Convention: Representation in Art and Science* (pp. 97–138). Dordrecht: Springer.

Giere, R. (1988). *Explaining Science*. Chicago: University of Chicago Press.

Giere, R. (1999a). *Science without Laws*. Chicago: University of Chicago Press.

Giere, R. (1999b). Using models to represent reality. In L. Magnani, N.J. Nersessian and P. Thagard (eds), *Model-Based Reasoning and Scientific Discovery* (pp. 41–57). Dordrecht: Kluwer Academic/Plenum Publishers.

Giere, R. (2004). How models are used to represent reality. *Philosophy of Science, 71*, 742–52.

Giere, R. (2009). Why scientific models should not be regarded as works of fiction. In M. Suárez (ed.), *Fictions in Science: Philosophical Essays on Modeling and Idealization* (pp. 248–58). London: Routledge.

Giere, R. (2010). An agent-based conception of models and scientific representation. *Synthese, 172*(2), 269–81.

'Glyptic formulae' (1867). *The Laboratory 1*, 78.

Godfrey-Smith, P. (2006). The strategy of model-based science. *Biology and Philosophy, 21*, 725–40.

Goldman, A. (2003). Representation in art. In J. Levinson (ed.), *The Oxford Handbook of Aesthetics* (pp. 192–210). New York: Oxford University Press.

Goodman, N. (1972). Seven strictures on similarity. In N. Goodman (ed.), *Problems and Projects* (pp. 23–32). New York: Bobbs-Merrill.

Goodman, N. (1976). *Languages of Art*. Indianapolis: Hackett.

Hacking, I. (1983). *Representing and Intervening*. Cambridge: Cambridge University Press.

Hofmann, A.W. (1865). On the combining power of atoms. *Chemical News*, 12, 166–90. Reprinted from *Proceedings of the Royal Institution of Great Britain*, 4, 301–30.

Hughes, R.I.G. (1997). Models and representation. *Philosophy of Science*, 64, S325–S336.

Hughes, R.I.G. (1999). The Ising model, computer simulation and universal physics. In M. Morgan and M. Morrison (eds), *Models as Mediators* (pp. 97–145). Cambridge: Cambridge University Press.

Kekulé, A. (1867). On some points of chemical philosophy. *Laboratory*, 1, 303–6.

Kekulé, A. (1963). Über die Constitution und die Metamorphosen der chemischen Verbindungen und über die chemische Natur des Kohlenstoffs. Trans. in O. Benfey (ed.), *Classics in the Theory of Chemical Combination* (pp. 115–31). New York: Dover (Reprinted from *Annalen der Chemie und Pharmacie* 1858, 106, 129–59).

Klein, U. (2003). *Experiments, Models, Paper Tools: Cultures of Organic Chemistry in the Nineteenth Century*. Stanford, California: Stanford University Press.

Knuuttila, T. (2005). Models, representation, and mediation. *Philosophy of Science*, 72, 1260–71.

Kolbe, H. (1877). 'Zeichen der Zeit: II', *Journal für praktische Chemie*, 123, 473–7.

Kroon, F. (1994). Make-believe and fictional reference. *The Journal of Aesthetics and Art Criticism*, 52(2), 207–14.

Laszlo, P. (2000). Playing with molecular models. *Hyle*, 6(1), 85–97.

Latour, B. (1990). Drawing things together. In M. Lynch and S. Woolgar (eds), *Representation in Scientific Practice* (pp. 19–68). Cambridge, Mass.: MIT Press.

Laymon, R. (1998). Idealizations. In E. Craig (ed.), *Routledge Encyclopedia of Philosophy*. London: Routledge. Retrieved from http://www.rep.routledge.com/article/Q048.

Le Bel, J.A. (1963). Sur des relations qui existent entre les formules atomiques des corps organiques et le pouvoir rotatoire de leurs dissolutions. Trans. in O. Benfey (ed.), *Classics in the Theory of Chemical Combination* (pp. 161–71). New York: Dover (Reprinted from *Bulletin de la societé chimique de France*, 1874, 22, 337–47).

Leslie, A.M. (1987). Pretense and representation: the origins of 'theory of mind'. *Psychological Review*, 94, 412–26.

Lewis, D. (1978). Truth in fiction. *American Philosophical Quarterly*, 15, 37–46.

Lossen, W. (1880). Über die Vertheilung der Atome im Raume. *Annalen der Chemie*, 204, 265–364.

Lynch, M. and Woolgar, S. (eds) (1990). *Representation in Scientific Practice*. Cambridge, Mass.: MIT Press.

Mazzolini, R.G. (2004). Plastic anatomies and artificial dissections. In S. de Chadarevian and N. Hopwood (eds), *Models: The Third Dimension of Science* (pp. 43–70). Stanford, California: Stanford University Press.

Meinel, C. (2004). Molecules and croquet balls. In S. de Chadarevian and N. Hopwood (eds), *Models: The Third Dimension of Science* (pp. 242–75). Stanford, California: Stanford University Press.

Meinong, A. (1960). The theory of objects. In R.M. Chisholm (ed.) *Realism and the Background of Phenomenology*. New York: Free Press (original work published 1904).

Morgan, M. and Boumans, M. (2004). Secrets hidden by two-dimensionality: the economy as a hydraulic machine. In S. de Chadarevian and N. Hopwood (eds), *Models: The Third Dimension of Science* (pp. 369–401). Stanford, California: Stanford University Press.

Morgan, M. and Morrison, M. (1999). Models as mediating instruments. In M. Morgan and M. Morrison (eds), *Models as Mediators* (pp. 10–37). Cambridge: Cambridge University Press.

Morrison, M. (1999). Models as autonomous agents. In M. Morgan and M. Morrison (eds), *Models as Mediators* (pp. 38–65). Cambridge: Cambridge University Press.

National Gallery (2008). Salisbury Cathedral from the Meadows, National Gallery Online. Retrieved from http://www.nationalgallery.org.uk/cgi-bin/WebObjects.dll/CollectionPublisher.woa/wa/work?workNumber=147.

Paternò, E. (1869). Intorno all'azione del percloruro di fosforo sul clorale, *Giornale di Scienze Naturali ed Economiche di Palermo*, 5, 117–22.

Pratt, A. (2006). DCU molecular model viewing galleries, School of Chemical Sciences, Dublin City University. Retrieved from http://www.dcu.ie/~pratta/jmgallery/index.htm.

Ramberg, P.J. (2001). Paper tools and fictional worlds: prediction, synthesis and auxiliary hypotheses in chemistry. In U. Klein (ed.), *Tools and Modes of Representation in the Laboratory Sciences* (pp. 61–78). Dordrecht: Kluwer.

Ramberg, P.J. (2003). *Chemical Structure, Spatial Arrangement: The Early History of Stereochemistry 1874–1914*. Aldershot, England: Ashgate.

Ramberg, P. and Somsen, G. (2001). The young J. H. van't Hoff: the background to the publication of his 1874 pamphlet on the tetrahedral carbon atom, together with a new English translation. *Annals of Science*, 58, 51–74.

Ramsay, O.B. (1974). Molecules in three-dimensions (II). *Chemistry*, 47(2), 6–11.

Ramsay, O.B. (1975). Molecular models in the early development of stereochemistry: I: The van't Hoff model. II: The Kekulé models and the Baeyer strain theory. In O.B. Ramsay (ed.), *van't Hoff-Le Bel Centennial* (pp. 74–96). Washington, D.C.: American Chemical Society.

Ramsay, O.B. (1981). *Stereochemistry*. London: Heyden.

Richardson, J.S., Richardson, D.C., Tweedy, N.B., Gernert, K.M., Quinn, T.P., Hecht, M.H. et al. (1992). Looking at proteins: representations, folding, packing, and design. *Biophysical Journal*, 63(5), 1186–209.

Rocke, A.J. (1981). 'Kekulé, Butlerov, and the historiography of the theory of chemical structure', *British Journal for the History of Science*, 14, 27–57.

Rocke, A.J. (1984). *Chemical Atomism in the Nineteenth Century: From Dalton to Cannizzaro*. Columbus, Ohio: Ohio State University Press.

Rocke, A.J. (1993). *The Quiet Revolution: Hermann Kolbe and the Science of Organic Chemistry*. Berkeley, California: University of California Press.

Rocke, A.J. (2010). *Image and Reality: Kekulé, Kopp, and the Scientific Imagination*. Chicago: University of Chicago Press.

Russell, B. (1956). On Denoting. In R.C. Marsh (ed.), *Logic and Knowledge* (pp. 41–56). London: George Allen and Unwin (original work published 1905).

Schier, F. (1986). *Deeper into Pictures*. Cambridge: Cambridge University Press.

Schorlemmer, C. (1894). *The Rise and Development of Organic Chemistry*, rev. edn. London: Macmillan.

Sheik, A. and Shaffer, J. (eds) (1979). *The Potential of Fantasy and Imagination*. New York: Brandon House.

Singer, D. and Singer, J. (1990). *The House of Make-believe*. Cambridge, Mass.: Harvard University Press.

Sklar, L. (2000). *Theory and Truth*. Oxford: Oxford University Press.

Snelders, H.A.M. (1975). J. A. Le Bel's stereochemical ideas compared with those of J. H. van't Hoff (1874). In O.B. Ramsay (ed.), *van't Hoff-Le Bel Centennial* (pp. 66–73). Washington, D.C.: American Chemical Society.

Sterrett, S.G. (2002). Physical Models and Fundamental Laws: Using One Piece of the World to Tell about Another. *Mind and Society*, 5, 51–66.

Stoker, B. (1994). *Dracula*. London: Penguin Popular Classics (original work published 1897).

Suárez, M. (1999). Theories, models, and representations. In L. Magnani, N.J. Nersessian and P. Thagard (eds) *Model-Based Reasoning and Scientific Discovery* (pp. 75–83). Dordrecht: Kluwer Academic/Plenum Publishers.

Suárez, M. (2003). Scientific representation: against similarity and isomorphism. *International Studies in the Philosophy of Science*, 17, 225–44.

Suárez, M. (2004). An inferential conception of scientific representation. *Philosophy of Science*, 71, 767–79.

Suárez, M. (ed.) (2009). *Fictions in Science: Philosophical Essays on Modeling and Idealization*. London: Routledge.

Suppes, P. (1960). A comparison of the meaning and use of models in mathematics and the empirical sciences. *Synthese*, 12, 287–300.

Thomson-Jones, M. (2007). Missing systems and the face value practice. Retrieved from http://philsci-archive.pitt.edu/archive/00003519. (Longer manuscript version of Thomson-Jones [2010].)

Thomson-Jones, M. (2010). Missing systems and the face value practice. *Synthese*, 172(2), 283–99.

Toon, A. (forthcoming). Similarity and scientific representation. To appear in *International Studies in the Philosophy of Science*.

Vaihinger, H. (1924). *The Philosophy of 'As If'*, trans. C.K. Ogden. London: Kegan Paul (original work published 1911).

van der Spek, T.M. (2006). Selling a theory: the role of molecular models in J.H. van't Hoff's stereochemistry theory. *Annals of Science*, 63, 157–77.

van Fraassen, B. (1980). *The Scientific Image*. Oxford: Clarendon.
van Fraassen, B. (2008). *Scientific Representation: Paradoxes of Perspective*. Oxford: Oxford University Press.
van Inwagen, P. (1977). Creatures of fiction. *American Philosophical Quarterly*, 14, 299–308.
van't Hoff, J.H. (1875). *La chimie dans l'espace*. Rotterdam: Bazendijk.
van't Hoff, J.H. (1877). *Die Lagerung der Atome im Raume*. Braunschweig: Vieweg.
van't Hoff, J.H. (1998). *The Arrangement of Atoms in Space*, trans. A. Eiloart. In D.M. Knight (ed.), *The Development of Chemistry 1789–1914 (Vol. X)*. London: Routledge (original work published in 1898).
van't Hoff, J.H. (2001). *A Proposal for Extending the Currently Employed Structural Formulae in Chemistry into Space, Together with a Related Remark on the Relationship between Optical Activating Power and Chemical Constitution*. Trans. in P. Ramberg and G. Somsen (eds), The young J.H. van't Hoff: the background to the publication of his 1874 pamphlet on the tetrahedral carbon atom, together with a new English translation. *Annals of Science*, 58, 51–74 (original work published 1874).
Walton, K. (1990). *Mimesis as Make-Believe*. Cambridge, Massachusetts: Harvard University Press.
Wells, H.G. (2005). *The War of the Worlds*. London: Penguin (original work published 1898).
Weisberg, M. (2007). Who is a modeler? *British Journal for Philosophy of Science*, 58(2), 207–33.
Weisberg, M. (forthcoming). *Simulation and Similarity: Using Models to Understand the World*. Oxford: Oxford University Press.
Weisberg, D.S. and Bloom, P. (2009). Young children separate multiple pretend worlds. *Developmental Science*, 12(5), 699–705.
Winsberg, E. (2009). A function for fictions: expanding the scope of science. In M. Suárez (ed.), *Fictions in Science: Philosophical Essays on Modeling and Idealization* (pp. 179–89). New York: Routledge.
Wislicenus, J. (1873) 'Über die optisch-active Milchsäure der Fleischflüssigkeit, die Paramilchsäure', *Annalen der Chemie*, 167, 302–46.
Wyman, E., Rakoczy, H. and Tomasello, M. (2009). Normativity and context in young children's pretend play. *Cognitive Development*, 24(2), 146–55.

Index

absolute isomers, 88–9
abstract objects, models as, 13–15, 19, 41–5, 47, 66–7
antirealism, *see* realism
authorised games, 36, 45, 49–50, 63

Baeyer, Adolf von, 94, 100, 105
Barton, Derek, 123
Berzelius, Jacob, 85
Berzelian formulas, 103–6
Biot, Jean-Baptiste, 88
Butlerov, Aleksandr, 85–6

Callender, Craig, 26–33, 63–5, 78–9
Cannizzaro, Stanislao, 102
Cartwright, Nancy, 9–10, 12, 14
characters, fictional
  accounts of, 15–17
  models as, 15–18, 19–20, 41–5, 47, 49, 53–60, 66–7, 82
chemical formulas, *see* formulas, chemical
chemical models, *see* models, chemical
chemical structure, 85–8
Cohen, Jonathan, 26–33, 63–5, 78–9
computer models, *see* models, computer
Constable, John, 20
Contessa, Gabriele, 18
Crick, Francis, 20, 23, 32–3, 66, 68–9
Crum Brown, Alexander, 87, 90, 95–7, 101
crystallography, 88–9

David, Jacques-Louis, 20
denotation, 31–3, 56, 58–9, 63–5, 76, 79–81

depiction
  and participation, 120–1
  and scientific representation, 20–3, 25–6, 30–3, 45, 67–8, 76–8
derivative accounts of representation, 25–6, 28–31, 33, 62, 63–5
description-fitting objects, 13–18, 41–5
  see also abstract objects, models as; characters, fictional
DNA model, 20, 23, 32–3, 66, 68–9
dolls, 117, 120–3, 125–9
Dracula, 15–16, 17, 36, 40–1, 45–6

entities
  abstract, *see* abstract objects, models as
  fictional, *see* characters, fictional
ether, models of, 22–3, 41, 75–83
experiments, imagined, 126–9
explanation, simulacrum account of, 10

face-value practice, 12–13, 15
fiction, works of
  biographical, 73–4
  discourse about, 48–53, 121
  historical, 73–4
  and models, 38–41, 42–3, 44, 45–7, 53–60, 69–75
  versus non-fiction, 35–7, 69–75
  Walton's theory of, 34–7, 69–71, 74–5, 120–1
  see also characters, fictional; fictionality; fictionalism; fictional truths
fictional characters, *see* characters, fictional

fictionalism, 71
fictionality
  compatibility with truth, 34–5, 74
  as defined in Walton's theory, 34–5
fictional truths, 34–5, 45–6
Fine, Arthur, 71
formulas, chemical, 87, *87*, 90, *91*, 93, 95, 100–7
Francoeur, Eric, 114, 123, 124, 129
Frigg, Roman, 18, 19–20, 24, 55–9

games, *see* make-believe, games of
Gay-Lussac, Joseph-Louis, 85
Gerhardt, Charles, 88
Giere, Ronald, 7, 13–15, 18, 19, 43–4, 50–3, 59–60, 66, 79, 82
glyptic formulae, 95, *96*
Godfrey-Smith, Peter, 18, 19, 55, 59–60, 82
Goodman, Nelson, 25, 31–2, 68, 77
Graves, Robert, 73

Hacking, Ian, 7
Hermann, Felix, 98
Hoff, J. H. van't, 84, 89–95, 97–107
Hofmann, August, 95, *96*, 101
Howard, Michael, 45
Hughes, R. I. G., 12, 22, 76, 79

I, Claudius, 73–4
iconic, chemical formulas as, 103–7
imagination, *see* experiments, imagined; make-believe, games of; participation, imaginative
imagined experiments, 126–9
indirect views of modelling, 17–18, 19–20, *19*, 41–5, *43*, 53–60, 66–7
inferential conception of scientific representation, 80–1
Inwagen, Peter van, 16
isomerism, 85–9
isomorphism accounts of representation, 79–80

Kekulé, August, 85, 86–7, 95–7, *96*, 100–2
Keneally, Thomas, 73
Klein, Ursula, 103–7
Kolbe, Hermann, 93–5, 101
Kroon, Frederick, 53

lactic acid, 88–9
laws, theoretical, 9–10, 12, 14–15
Laymon, Ronald, 11
learning
  in children's games, 127–8
  in modelling, 10–13, 45–7, 65–7, 126–9
  two-stage view of, 67, 127–8
Le Bel, J. A. 91–3
Lewis, David, 42–3
Lossen, Wilhelm, 94–5

make-believe, games of, 34–5, 108, 120–3, 125–9
  see also participation, imaginative
matching, 37–8, 40, 66
Meinel, Christoph, 84, 100–1, 106, 107
Meinong, Alexius, 16
Milkmaid, The, 77
missing systems, 10–13, 17, 55
model-representation, 20–6, 28, 31, 32–3, 61–7, 77–9, 81–2
models
  as abstract objects, *see* abstract objects, models as
  accuracy of, 23–5, 50–3, 65–7, 116–17
  chemical, 7, 76, 95–107, 109–10, 117–19, 121–9
  computer, 110, 113, 115–16, 119, 123, 126, 129
  of DNA, 20, 23, 32–3, 66, 68–9
  of the economy, *see* Phillips machine
  of the ether, 22–3, 41, 75–83
  as fictional characters, *see* characters, fictional
  learning with, *see* learning, in modelling

manipulation of, 114–16, 121–3, 126–9
molecular, *see* models, chemical
ontology of, 10–18, 41–5, 57–60
physical, 6–7, 37–8, 95–102, 109–10, 111–15, 117–19, 121–9
predator-prey, 12, 77
scale, 7, 11, 13, 37–8, 63, 105, 127
and scientific realism, 1–2, 23–5, 65–7, 116–17
scientists' talk about, 10–13, 17, 48–53, 111–13, 121
of simple harmonic oscillator, 7–8, 11–13, 15, 38–40, 42–3, 44–5, 46–7, 48–53, 77
of solar system, 9, 55–6
theoretical, 7–18, 19–20, 38–41, 41–5, 45–7, 48–53, 54–60, 65–7, 71–5, 76
and theory, 10, 100–2
three-dimensional *see* models, physical
without objects, 22–3, 40–1, 54–5, 75–82, 106
see also model-representation; model-systems
model-systems, 13, 15, 17–18, 19–20, 43–4, 47, 48, 54–60, 66–7, 82
see also models, ontology of
molecular models, *see* models, chemical
Morgan, Mary, 10
Morrison, Margaret, 10, 12

Napoleon Crossing the Saint Bernard, 20, 51, 77
Newtonian model of solar system, 9, 55–6
nonderivative, *see* derivative accounts of representation
nonexistent objects, models of, *see* models, without objects

objects
abstract, *see* abstract objects, models as
fictional, *see* characters, fictional
of representations, 36–7
optical activity, 88–9, 90–3

paper tools, 103, 107
participation, imaginative
in children's games, 120–3, 125–9
and depiction, 120–1, 123–5
with fiction, 48–55, 120–1
in modelling, 48–53, 111–29
tactile, 119–23, 125–9
verbal, 48–53, 121
visual, 119–23, 123–5
Pasteur, Louis, 88, 90–2
Paternò, Emmanuele, 102
Phillips machine, 7, 38, 63, 69, 77, 78–9, 81, 105
physical models, *see* models, physical
pictures, *see* depiction
Plato, 26
play, *see* make-believe, games of
prepared descriptions, 10
pretence, 48–53, 116–17, 118, 120–1
principles of generation, 34, 36, 37–9, 45–7, 65, 68–9, 105, 106, 118–19
principles of direct generation, 46
principles of implication, 46
props
definition of, 34–7
models as, 37–41, 61–3, 74–5, 81–2, 104–5, 117–19
scientists as, 119–29

Ramberg, Peter, 84, 92, 95, 103–4
realism
about fictional entities, 16–17
scientific, 1–2, 23–5, 65–7, 116–17
reference, *see* denotation
reflexive props
definition of, 36
models as, 40, 117, 119
scientists as, 119–29

representation
  accuracy of, 23–5, 50–3, 65–7, 116–17
  and denotation, *see* denotation
  derivative and nonderivative accounts of, 25–6, 28–31, 33, 62, 63–5
  make-believe account of, 37–41, 61–9, 81–2
  and misrepresentation, 23–5, 65–7
  of non-existent entities, *see* models, without objects
  pictorial, *see* depiction
  and realism, 1–2, 23–5, 65–7, 116–17
  as a relation, 22–3, 79–82
  similarity accounts of, *see* similarity accounts of representation
  and stipulation, 29, 31–3, 63–5
  Walton's definition of, 35–7
representation-as, 37, 64–5
resemblance theories of depiction, 25–6, 32, 67–9, 124
  see also similarity accounts of scientific representation
richness, 124
Rocke, Alan, 84, 107
Russell, Bertrand, 16

St Paul's Cathedral, 39
Salisbury Cathedral from the Meadows, 20
Schorlemmer, Carl, 102
scientific representation, *see* representation
semantic view of theories, 134
semiotics, Peircean, 103
silogen atoms, 72–3
similarity accounts of representation, 19–20, 60, 66, 67–9, 79–80
simple harmonic oscillator, *see* models, of simple harmonic oscillator
solar system, model of, 9, 55–6
statues, 37–8, 121
stereochemistry
  development of, 84–5, 88–95

  models in, 97–107
stipulation and scientific representation, 29, 31–3, 63–5
Stoker, Bram, 55
structural theory, 85–8
Suárez, Mauricio, 23–4, 68, 71, 80–1
symbolic, chemical formulas as, 103–7

talk about models, 10–13, 17, 48–53, 111–13, 121
tartaric acid, 88, 90–1
tetrahedral carbon atom, *see* stereochemistry
theoretical hypotheses, 50–3, 79
theoretical models, *see* models, theoretical
Thomson-Jones, Martin, 11–12, 13, 15, 18, 135, 136
Tolstoy, Leo, 74
toys, 107
  see also dolls
truth
  and accuracy, 65–7
  compatibility with fiction, 34–5, 74
truths, fictional, 34–5, 45–6

unofficial games, 36, 51–3

Vaihinger, Hans, 71–2
Van't Hoff, J. H., *see* Hoff, J. H. van't
Vermeer, Johannes, 77

Walton, Kendall
  on fiction and nonfiction, 35–6, 69–71, 74–5
  pretence account of discourse about fiction, 49–50, 51–3, 121
  theory of depiction, 120–1
  theory of representation, 34–7, 81–2
War and Peace, 36–7, 74
War of the Worlds, The, 37, 39, 42
Watson, James, 20, 23, 32–3, 66, 68–9
Wells, H. G., 39
Winsberg, Eric, 72–3
Wislicenus, Johannes, 88–9, 94, 97–8, 106